Spotlight on Standards!
Interactive Science Content Reader

Printed in the United States of America

ISBN-13: 978-0-15-365363-6
ISBN-10: 0-15-365363-9

10 11 12 13 14 0877 16 15 14 13

4500425971

Standard Set 1 Physical Science

Unit 1 Energy and Matter

1 Energy and matter have multiple forms and can be changed from one form to another.

Standard Set 2 Physical Science

Unit 2 Light

2 Light has a source and travels in a direction.

Standard Set 3 Life Science

Unit 3 Adaptations

3 Adaptations in physical structure or behavior may improve an organism's chance for survival.

Standard Set 4 Earth Science

Unit 4 Patterns in the Sky

4 Objects in the sky move in regular and predictable patterns.

Energy and Matter

In this unit, you will learn about how water moves from Earth to air and back again, how we get the water we need, and ways to conserve water. What do you know about these topics? What questions do you have?

Thinking Ahead

What are some sources of energy? Draw a picture to show what you think.

How does a liquid turn to a gas? Draw a picture to show what you think.

Does your drawing show a chemical or physical change?

____

Circle the name of the object you think is smaller.

atom **grain of sand**

Write a question you have about energy and matter.

____

____

____

Recording What You Learn

On this page, record what you learn as you read the unit.

Lesson 1

What are two types of energy that come from the sun?

_______________ _______________

Lessons 2 and 3

How does a machine move electricity and use stored energy? Draw a picture.

Lessons 4, 5, and 6

Explain the difference between a physical and chemical change.

____

____

____

____

Lesson 7

Name a substance that is an element. Name a substance that is not.

____

____

California Standards in This Lesson

 **1.a** *Students know* energy comes from the Sun to Earth in the form of light.

Vocabulary Activity

energy

light

heat

Use the vocabulary words above to answer the following questions.

1. What does energy do? Give an example.

____

____

2. How are two forms of energy from the sun used?

____

____

____

Lesson 1

VOCABULARY
energy
light
heat

Where Does Energy on Earth Come From?

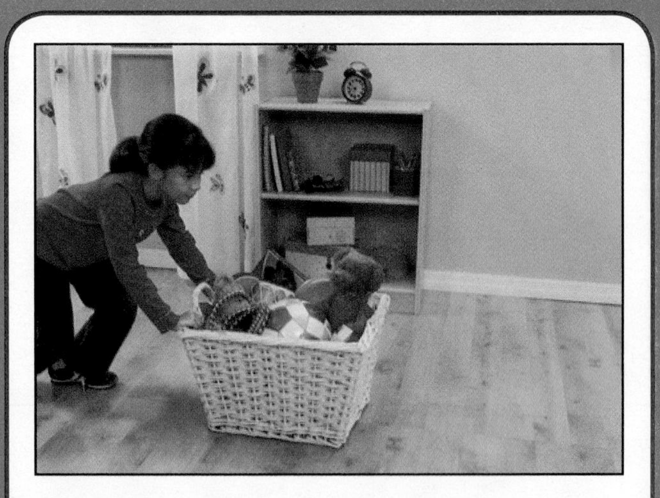

Energy makes things change. The girl uses energy to push the basket across the floor.

Light is energy that you can see. The light from the flashlight shines on the wall.

Heat is also a form of energy. The heat from this stove can cook food.

Hands-On Activity
The Sun's Light

1. Sunlight is a kind of energy that makes plants grow. Look at healthy trees and plants. Find trees or plants that have not gotten enough energy from the sun.

2. How are those trees and plants the same?

____

____

3. How are they different?

____

____

____

____

Concept Check

1. The **Main Idea** on these two pages is Energy causes changes that people can use. **Details** tell more about the main idea. Find details about how energy causes change and how people can use these changes. Underline two of them.
2. Where does the energy come from that makes the windsurfer's board move?

____

3. What change happens to the surfer and the surfboard?

____

What Energy Does

We need energy to do everything. The surfer uses energy to hold on to the sail and move it.

The waves carry energy that makes the board move. **Energy** causes changes. The windsurfer uses energy from himself and from the water and the wind to surf.

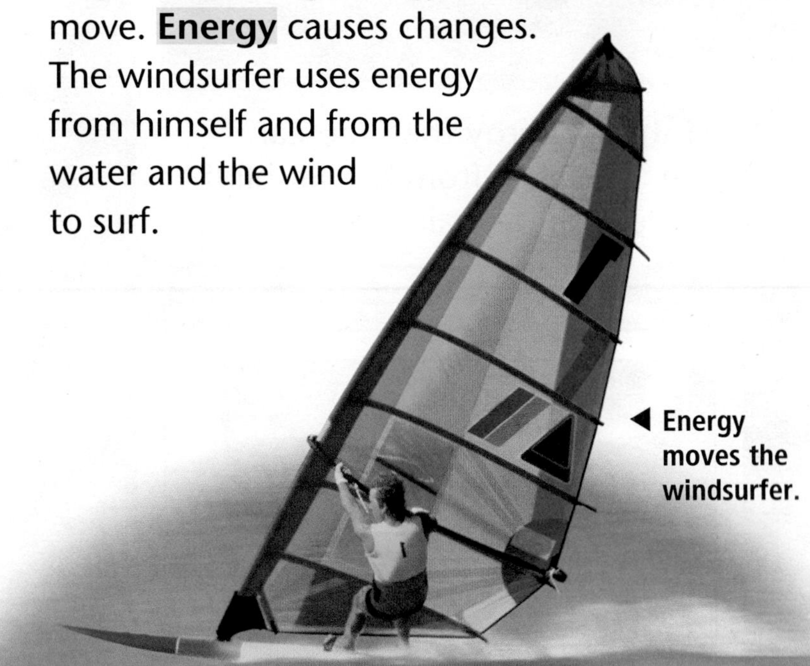

◀ Energy moves the windsurfer.

Energy makes things happen. The moving water has the energy to turn the water wheel. Every action needs energy. You use energy to go to school, to read a book, to run, and to play ball.

✓Concept Check

1. What are some ways you use energy?

____

2. Every action needs ______________.

3. Draw a picture showing how water moves a water wheel. Draw an arrow to show the direction of the water.

✓Concept Check

1. The **Main Idea** on these two pages is There are many sources of energy. **Details** tell more about the main idea. Find details about some sources of energy. Underline two of them.

2. Draw a picture showing energy provided by the sun.

Some Sources of Energy

There are many sources of energy. Food is a source of energy. When you eat food, you give your body energy to grow. Energy from wind makes flags wave.

Light is energy from the sun that you can see. Heat energy from the sun makes you feel warm.

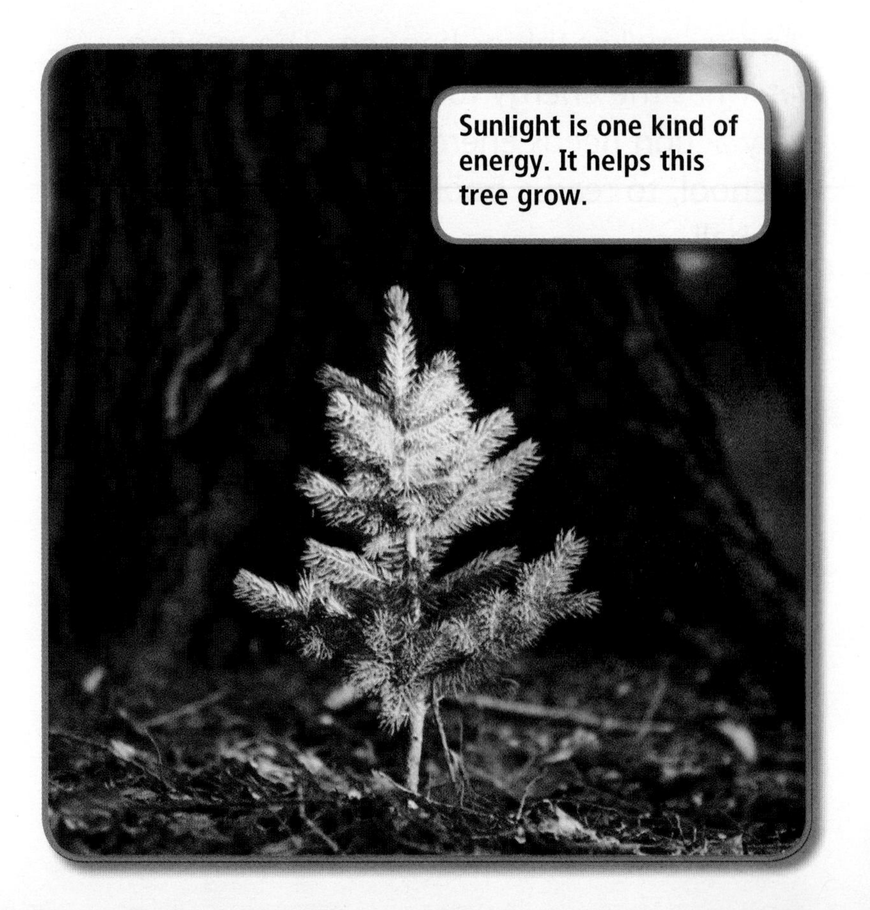

Sunlight is one kind of energy. It helps this tree grow.

Light from the Sun

The sun's energy is very important to all living things. Sunlight is important to people, plants, and animals.

Plants need sunlight to grow. People and some animals eat plants. Food gives us energy that lets us do many things.

This house has a solar panel. It changes the energy from sunlight to energy that machines in the house can use.

The solar panels face the sun to take in energy.

✓Concept Check

1. Why do we need sunlight?

____

____

____

2. Finish the sentence.

   Solar panels change the __________ from __________ to energy that machines in the house can use.

3. How are plants and solar panels alike?

____

____

✓Concept Check

1. The **Main Idea** on these two pages is Most of Earth's energy comes from the sun. **Details** tell more about the main idea. Find details about how the sun's energy changes things on Earth. Underline two of them.

2. What is one form of energy that the sun gives us?

____

3. Draw an example of how people use heat from the sun.

Heat from the Sun

The sun also gives Earth heat. **Heat** is energy that warms an object. All living things need heat. Without the sun's heat, Earth would be so cold that we could not live here.

The sun keeps us warm.

Most of Earth's energy comes from the sun. The sun's energy makes the wind blow. Wind energy can be used to make electricity. The energy in coal and oil came from the sun. We burn these fuels to get light and heat.

The sun gives us energy to make the electricity that lights this sign.

Lesson Review

Fill in the blanks to give Details about energy and how it is used.

1. Energy causes ____________.
2. A form of energy that you can see is ____________.
3. ____________ and ____________ come from the sun and give all living things energy.
4. A form of energy that warms objects is ____________.

California Standards in This Lesson

 **1.b** *Students know* sources of stored energy take many forms, such as food, fuel, and batteries.

 **1.c** *Students know* machines and living things convert stored energy to motion and heat.

Vocabulary Activity

fuel

battery

electricity

Use the vocabulary words above to answer the following questions.

1. What is the difference between energy from fuel and energy from batteries?

____

____

2. How do machines in your house get energy?

____

____

Lesson 2

VOCABULARY
fuel
battery
electricity

How Is Energy Stored and Used?

Fuel is something that is burned to produce energy. The girl is putting fuel into the car. The fuel gives the car energy to move.

These batteries have chemicals stored inside. This is chemical energy. A **battery** changes chemical energy into electrical energy.

Electricity is energy that can move through wires. When we put wires in a wall, electricity moves through them.

Hands-On Activity
Explain and Observe

1. Help an adult make a meal. Write down the names of two of the foods in the meal. Explain how each of those foods stores energy.

____

____

____

2. Walk around your home. What kinds of objects use batteries and electricity to work? List three of each.

Batteries	Electricity
_______	_______
_______	_______
_______	_______

✓Concept Check

1. **Sequence** is the order in which things happen. Sometimes words such as *before*, *first*, *next*, *after*, and *then* help you understand the sequence. Underline the words on these two pages that tell about sequence.

2. What must happen to fuel before we can use its energy?

____

3. Use page 12 to finish the table.

Types of Fuel	Where They Come From	What They Are Used For
Coal, Oil, and Natural Gas	____________ ____________ ____________	____________ ____________ ____________

Energy from Fuels

Fuel is something that is burned to get the energy stored in it. Coal, oil, and natural gas are fuels. These fuels come from living things that died long ago.

Coal, oil, and gas store energy. When they are burned, they give up their energy. We use the energy to move things, such as a car, or to heat things, such as our food.

What must happen to fuel before we can use its energy?

When the gas burns, its stored energy changes into heat that cooks our food.

Energy from Food

People and animals get the energy they need from food. First, plants take in the sun's energy. They store it in their fruit and other parts. Then, people eat the plant parts. This family is eating. The food gives each person energy to move and grow.

When you eat, your body breaks food down into tiny parts that are too small to see. Some of these parts travel to your muscles. They give your muscles energy to move your body.

✓Concept Check

1. Write the numbers 1, 2, 3, and 4 to show the sequence.

_____ **Eating food gives each person energy to move and grow.**

_____ **Plants store energy in their fruit and other parts.**

_____ **Plants take in the sun's energy.**

_____ **People eat the plant parts.**

2. Draw a picture that shows how you got energy today.

✓Concept Check

1. **Sequence** is the order in which things happen. Sometimes words such as *when*, *after*, and *then* help you understand the sequence. Underline the words on these two pages that tell about sequence.
2. What happens after a machine gets energy?

____

3. Use pages 11 and 15 to finish the chart.

	Where does the electricity come from?	What is the electricity used for?
Wall outlet		
Batteries		

Energy from Machines and Batteries

People use energy to make machines do work. A car needs gasoline to move. Energy is stored in the gasoline. When the gasoline burns, its stored energy is released. This energy then turns the wheels to move the car.

You use energy to move the pedals of a bike. The pedals are on a wheel that turns bigger wheels. This makes the bike move.

Burning gasoline releases energy that turns the wheels of the car.

A **battery** holds chemicals. It stores energy in these chemicals. When the battery is used, the chemical energy changes to electricity.

Electricity is energy that can light a flashlight or move a toy car. If you take the batteries out of the toy car, it will not be able to move.

Lesson Review

Complete these Sequence sentences.

1. Oil, gas, and ___________ are fuels that formed long ago.
2. When a ___________ is burned, its stored energy is released.
3. After people eat ___________, their bodies turn it into energy.
4. When something that uses a battery is turned on, chemical energy changes to ___________.

California Standards in This Lesson

 **1.d** *Students know* energy can be carried from one place to another by waves [such as water waves and sound waves], by electric current, and by moving objects.

Vocabulary Activity

wave

vibrations

friction

Use the vocabulary words above to complete the following sentences.

1. Objects slow down because of ________.
2. A ____________ is something that carries energy.
3. When you hear a sound, it was caused by ____________.

Lesson 3

VOCABULARY
wave
vibrations
friction

How Does Energy Move?

The raindrop makes rings in the water. Each ring is like a wave. A **wave** carries energy across the water.

The alarm clock rings. The air moves quickly back and forth. We hear these **vibrations** as sound.

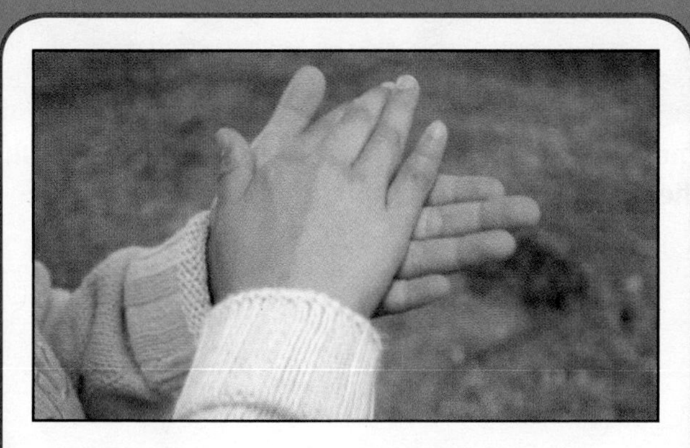

Rubbing your hands together causes friction. **Friction** can slow the movement of objects that touch. Friction can make your hands warm.

Hands-On Activity Waves

1. Fill a bowl partway with water. Put different objects in the water. With your hand, make little waves in the water. What happens to the objects?

____

2. Now make bigger waves. What happens to the objects now?

____

3. Put your hand on your throat. Say this sentence in a normal voice: "My voice makes sound waves when I talk." Then say it loud! Write how your throat felt each time you said the sentence.

____

____

____

____

✓Concept Check

1. You **Compare** when you look at how things are alike. You **Contrast** when you look at how things are different. Contrast water waves and light waves.

How are water waves and light waves different?

___

___

___

2. All of the blanks in the sentences below can be filled in with the same word. Circle the word.

The ________ carry the wind's energy.

Light and heat from the sun move to Earth in ________.

These ________ move up and down, like ________in water.

wind **electricity** **waves**

Waves Carry Energy

Wind blowing across water makes waves. The waves carry the wind's energy. Waves move up and down. They also move anything that is on the water up and down.

Light and heat from the sun move to Earth in waves. These waves move up and down, like waves in water. They also carry energy. You cannot see these waves.

▲ **The waves carry energy across the water. This causes the penguins to move up and down.**

Heat from the sun will cause the snowman to melt.

An earthquake makes the ground move. Energy from the earthquake moves through the ground in waves.

Houses on the ground cannot move. Sometimes, they fall down in an earthquake.

✓Concept Check

1. Classify each type of wave as visible or invisible.

Wave	Visible or Invisible?
Water	
Light	
Heat	
Earthquake	

2. What kind of energy makes a snowman melt?

____

3. Why do earthquakes do so much damage to buildings?

____

____

✓Concept Check

1. You **Compare** when you look at how things are alike. You **Contrast** when you look at how things are different. You can compare and contrast sound waves and light waves.

How are sound waves and light waves the same?	How are sound waves and light waves different?

2. Circle the verb that means "move quickly back and forth" each time it appears on these two pages.

3. Underline the noun that means "quick movements back and forth" each time it appears on these two pages.

Sound Waves

An object that *vibrates* moves quickly back and forth. These quick movements are **vibrations**. The vibrations produce sound waves. Sound waves do not move up and down like water waves.

The strings on a harp vibrate, making sound waves. The sound waves move back and forth. The waves carry sound energy away from the strings.

◀ The strings are different sizes. They vibrate differently to make music.

Touch your throat when you talk. Feel the vibrations. Your voice makes sound waves when you talk. The singing bird also makes sound waves.

The *eardrum* is a thin sheet inside the ear. When sound waves hit the eardrum, it vibrates. The vibrations travel through the ear. The ear sends a message to the brain. Your brain tells you what you hear.

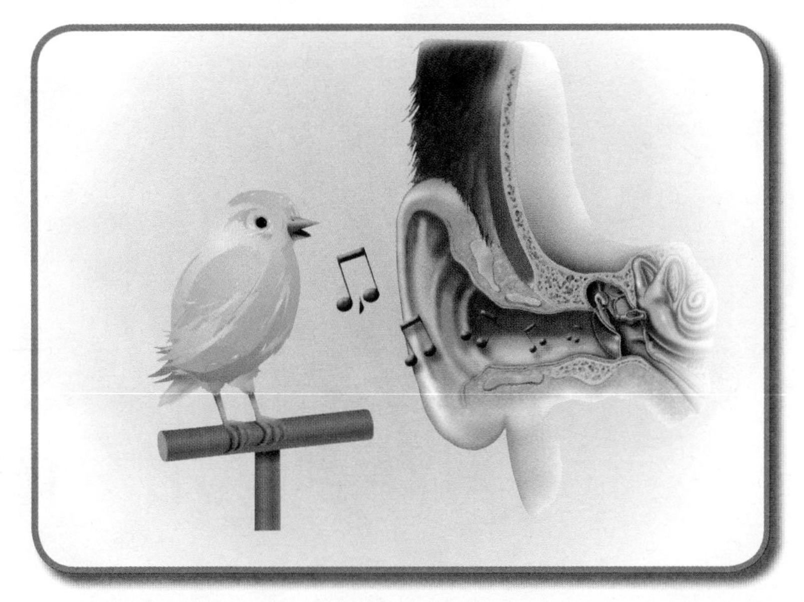

▲ **The bird's song moves through the air as sound waves. Your eardrum vibrates when the sound waves reach it.**

✓Concept Check

1. Write 1, 2, 3, and 4 to put the events in sequence.

Your brain tells you what you hear.	
Vibrations travel through the ear.	
Sound waves hit the eardrum.	
The ear sends message to the brain.	

2. How do you think an eardrum is like a musical drum?

____

____

____

✓Concept Check

1. You **Compare** when you look at how things are alike. You **Contrast** when you look at how things are different. You can compare and contrast other ways energy moves by filling in the chart below.

How are wind farms and rubbing hands the same?	How are wind farms and rubbing hands different?

2. Write the steps that tell how wind makes a television work in your home.

1. ________________________________

2. ________________________________

3. ________________________________

Other Ways Energy Moves

Electricity is a kind of energy that does not travel in waves. How does electricity move?

Energy stations change the energy of wind, water, or fuel into electricity. Then the electricity travels through wires to your home.

An energy station sends electricity through wires to your home. ▼

▲ A wind farm is an energy station that uses the energy of the wind to produce electricity.

If you rub your hands together, you feel friction. **Friction** slows down objects that are touching each other. The objects get warm. Friction changes the energy of moving objects to heat energy.

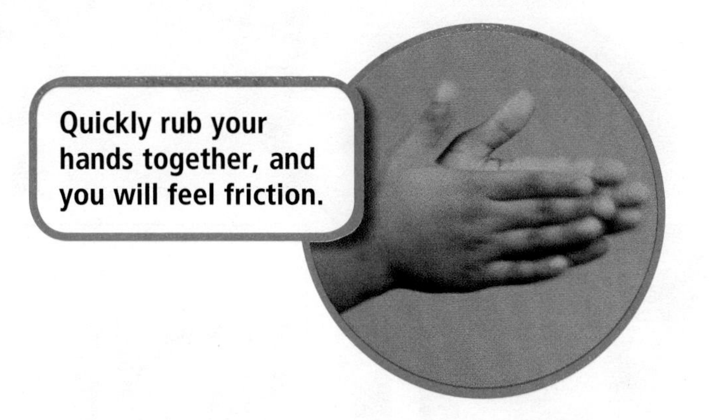
Quickly rub your hands together, and you will feel friction.

Lesson Review

Complete these sentences that Compare and Contrast.

1. Both water waves and earthquake waves move ___________ and ___________.
2. Sound waves and water waves both carry ___________.
3. Light travels in waves, but ___________ travels through wires.
4. Water waves move up and down, but ___________ waves do not.
5. Light, heat, and sound energy travel in ___________. ___________ is a kind of energy that does not travel in waves.

California Standards in This Lesson

 **1.e** *Students know* matter has three forms: solid, liquid, and gas.

Vocabulary Activity

Melanie filled out the chart below with some of the vocabulary words. Check her answers. If she's right, draw a star. If she's wrong, write the correct word on the line.

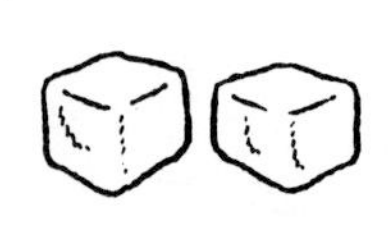

solid

____

gas

____

liquid

____

Lesson 4

What Is Matter?

VOCABULARY
matter
solid
liquid
gas

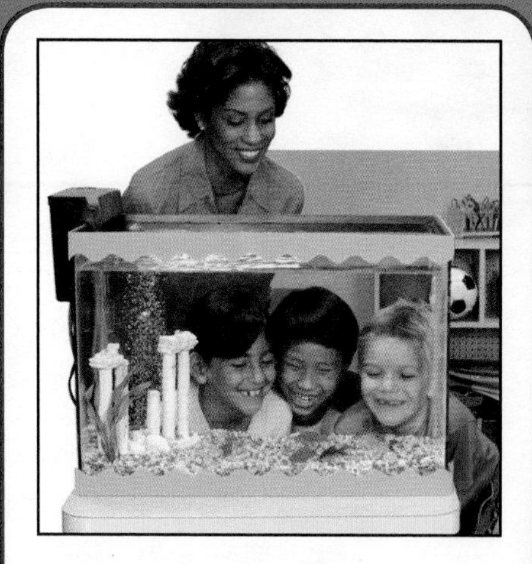

Everything in this picture is **matter**. The people, the fish, the water, and the air are all matter.

The containers hold jellybeans, pasta, colored sand, and marbles. Each has a definite shape and volume. Each is a **solid**.

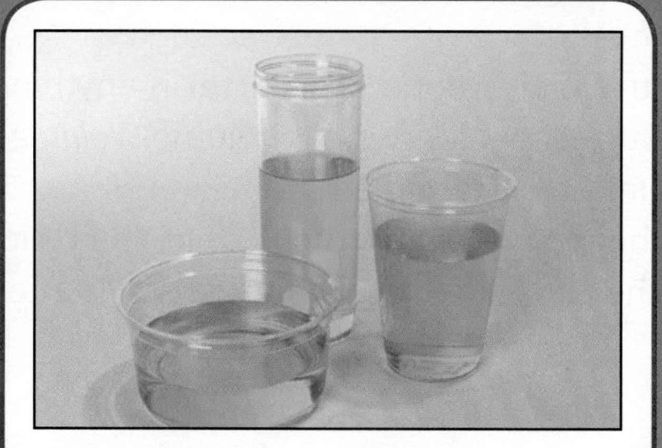

Each container holds a liquid. A **liquid** has a set volume, but it takes the shape of the container that holds it.

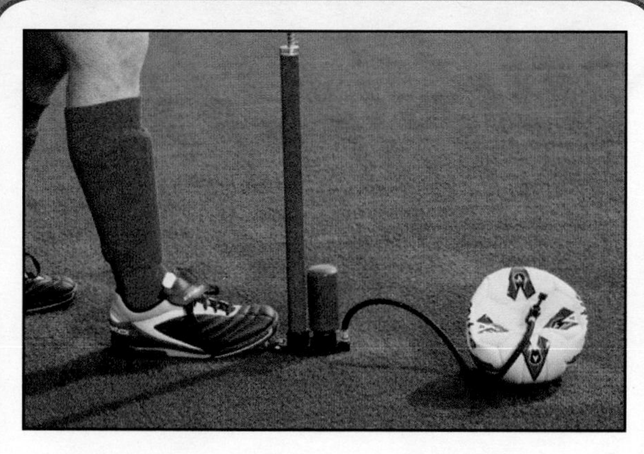

The soccer player is pumping air into the ball. Air is a **gas**. The shape and volume of a gas can change.

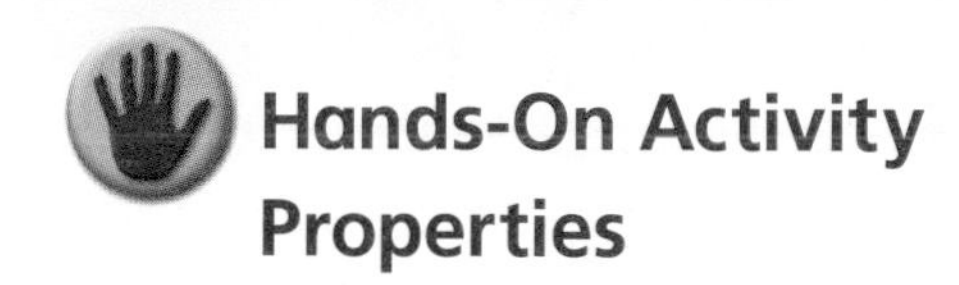

Hands-On Activity
Properties

1. List the names of solids, liquids, and gases that you see every day. When you have named three of each kind of matter, share the list with a friend.

Solids	Liquids	Gases
______	______	______
______	______	______
______	______	______

2. Look around your home and pick an object. Describe its properties. Use sight, touch, smell, and sound.

My object: ______________________

Sight: ______________________

Touch: ______________________

Smell: ______________________

Sound: ______________________

✓Concept Check

1. The **Main Idea** on these two pages is Everything around us is matter that takes up space. **Details** tell more about the main idea. Find details about how everything around us is matter that takes up space. Underline two of them.

2. What is matter?

____

3. What matter do you see in the picture on page 26?

____

Matter

Matter is all around us. **Matter** is anything that takes up space. Matter has volume. *Volume* is the amount of space something takes up.

In the picture below, the people, the balloons, and the air inside the balloons are all matter.

Forms of Matter

Matter has different forms, or *states*. Matter can be a solid, a liquid, or a gas.

A **solid** is one form of matter. The volume and shape of a solid do not change when you move it from place to place.

The girl and her book are both solids. When the girl moves, her shape and volume do not change. The book's shape and volume also stay the same.

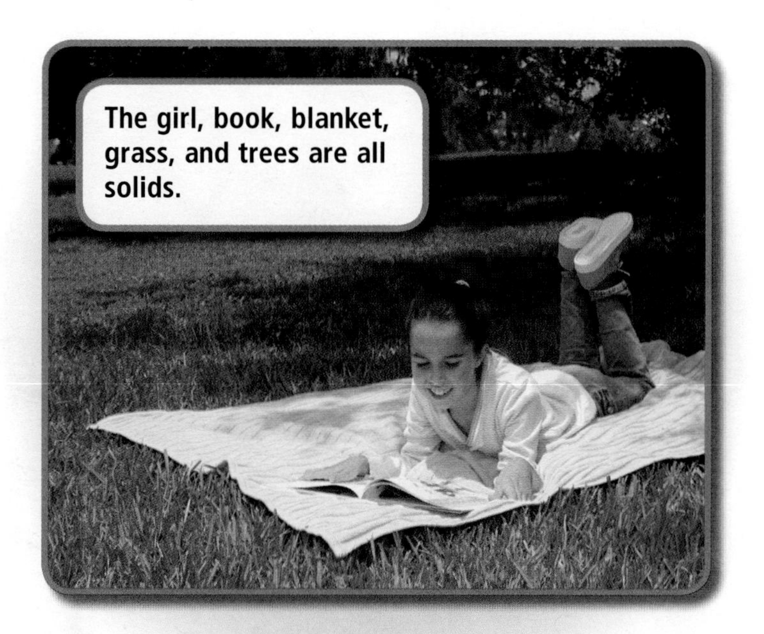

The girl, book, blanket, grass, and trees are all solids.

✓Concept Check

1. What are three forms of matter?

___

2. What happens to the volume and shape of a solid when you move it from place to place?

___

3. If the matter is a solid, check SOLID. If the matter is not a solid, check NOT SOLID.

Matter	SOLID	NOT SOLID
eraser		
milk		
air		
juice		
hamster		

✓Concept Check

1. The **Main Idea** on these two pages is Liquid and gas are states of matter. **Details** tell more about the main idea. Find details about how liquid and gas are states of matter. Underline one detail that tells about liquid. Underline one detail that tells about gas.

2. What is a liquid?

____

____

3. Complete the chart to tell what would happen if the water in the fish tank were poured into jars or buckets. Write the words **change** or **no change.**

Total Volume	Shape	Amount	State of Matter

Liquids

Water is a **liquid**. The amount, or volume, of a liquid does not change when you pour it from one container into another. The liquid's shape does change.

Look at the picture. The water takes the shape of the fish tank. If all of this water is poured into jars or buckets, its total volume will not change. The water's shape will change. The water will take the shape of the new containers.

The water in the tank has a definite volume but no definite shape.

If you pop the balloon, the air will go out of it and mix with the surrounding air.

Gases

A **gas** is matter that has no *definite* shape. This means that the shape of a gas changes. The air inside a balloon is a gas. The gas takes the shape of the balloon. If you pop the balloon, the air goes out of it. The air takes the shape of the space the balloon is in.

A gas has no definite volume. When the balloon pops, the air that was inside it spreads out. It becomes part of the air around the balloon. Its volume has changed.

The airplane and the mountains are solids. The water is a liquid. The air is a gas.

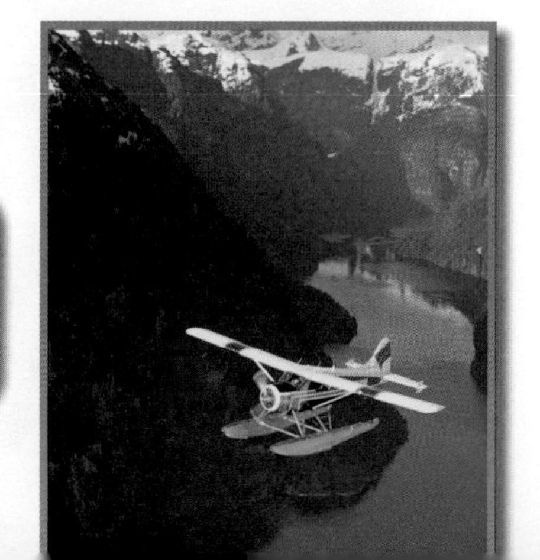

✓Concept Check

1. What are gases?

____

2. When you pop a balloon, what happens to the air inside it?

____

3. Write the names of three things that have gas inside them.

   1. _______________
   2. _______________
   3. _______________

4. Draw a line between the state of matter and its properties.

solid	no definite shape, no definite volume
gas	no definite shape, definite volume
liquid	definite shape, definite volume

✓Concept Check

1. The **Main Idea** on these two pages is Matter has properties. **Details** tell more about the main idea. Find details about the properties of matter. Underline two of them.

2. Name three properties of matter.

____

____

3. Use page 31 to complete the chart.

Tool	Measurement
Ruler	
Measuring Cup	
Balance	

Properties of Matter

Matter has properties. A *property* tells what something is like. One property you can see is color. Another property tells how an object feels when you touch it. Other properties tell how something tastes, smells, or sounds. You can say that milk is white, a cat's fur is soft, and a pan makes noise when you hit it. You use your senses to tell about these properties.

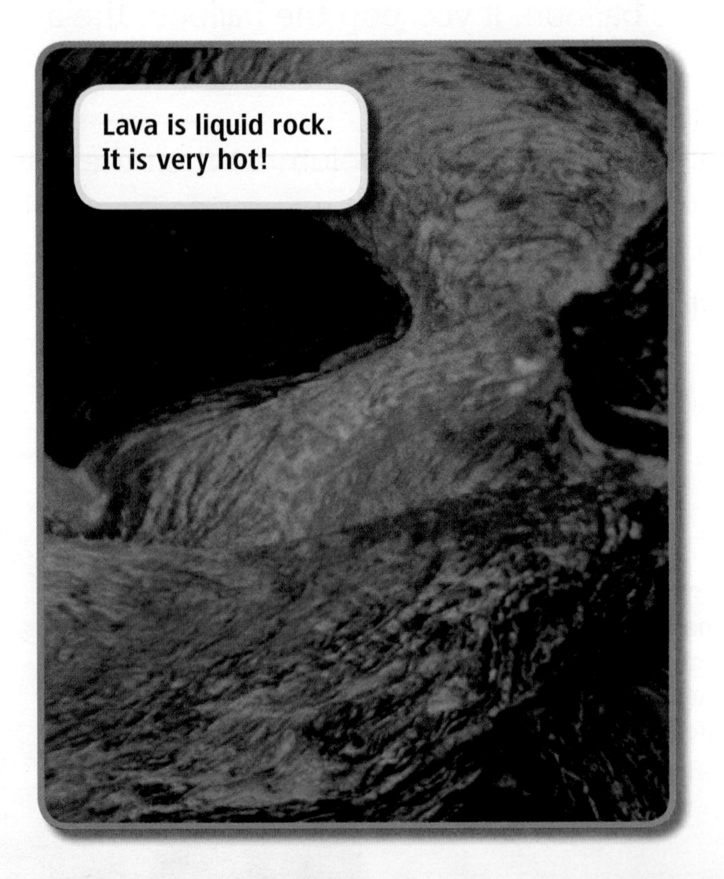
Lava is liquid rock. It is very hot!

You can also use tools to tell about matter. You can measure length with a ruler. You can use a measuring cup to find the volume of a liquid. You can use a balance to measure a property called mass. *Mass* is the amount of matter in an object.

The mass of the truck and the pebbles is 64 grams. ▶

Lesson Review

Complete the sentence about the Main Idea of the lesson.

1. Anything that takes up space is ____________.

Fill in the Detail statements about matter.

2. A ____________ has a definite volume and shape.
3. A ____________ has a definite volume but no definite shape.
4. A ____________ has no definite shape or volume.
5. Matter has ____________.

California Standards in This Lesson

 **1.e** *Students know* matter has three forms: solid, liquid, and gas.

 **1.f** *Students know* evaporation and melting are changes that occur when the objects are heated.

Vocabulary Activity

Read the descriptions of the vocabulary words on pages 32 and 33. Then answer the questions.

1. When would you see something melting?

____

____

2. When would something evaporate?

____

____

Lesson 5

What Causes Matter to Change State?

VOCABULARY
melting
evaporation

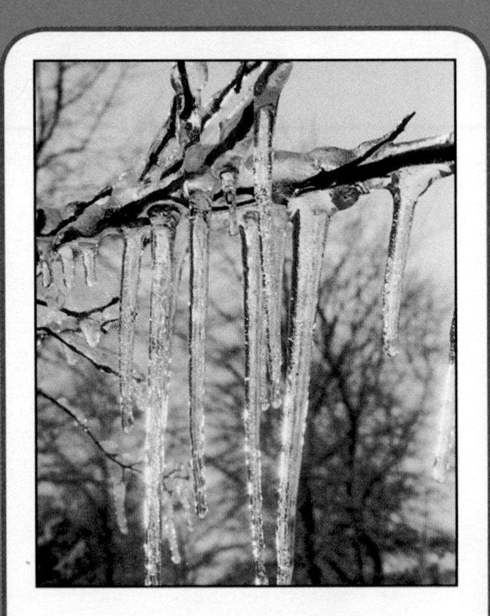

The ice is **melting**. It is changing from a solid to a liquid.

The sun is shining. The wet footprints are drying up. The sun is causing the water to evaporate. During **evaporation**, a liquid changes to a gas.

Hands-On Activity
Two States of Matter

1. Fill a small paper cup with water. Draw a picture of what it looks like in the box. Then label its state of matter.

State of matter: ____________________

2. Put the cup of water in the freezer. Wait 2 hours, then take the cup out of the freezer. Squeeze the ice in the cup into a bowl. Draw a picture of what it looks like in the box. Label its state of matter.

State of matter: ____________________

✓Concept Check

1. A **Cause** is something that makes another thing happen. An **Effect** is the thing that happens. Look for a cause and effect on these two pages. Underline a cause and circle its effect.
2. What causes a solid to melt? ___________
3. Read the statements below. Then tell whether they are **True** or **False.**

Heat causes particles to move around freely.

____

Melting is the change of a liquid to a gas.

____

All matter is made up of small pieces, or particles. ___________

Melting

Matter can change form, or state, when it is made cooler or warmer. Why? All matter is made up of small pieces, or *particles*.

Suppose you put some ice cubes into a pan of hot water. Heat causes the particles in the ice cubes to move around freely. The ice melts and becomes liquid water. **Melting** is the change of a solid to a liquid.

Iron is a metal. It melts at a very high temperature.

The boy's clothes, skin, and hair are wet.

Evaporation

In the picture above, the boy is wet. In the picture below, the boy is dry.

As the boy sits outside, heat energy from the sun causes the particles of liquid water to move faster and faster. The particles go into the air. The liquid water evaporates, or becomes a gas. **Evaporation** is the change of a liquid to a gas.

The boy is now dry. The sun's energy caused the liquid water to become water vapor, a gas.

✓Concept Check

1. What causes evaporation?

____

2. Number the statements below to show how evaporation makes the wet boy dry.

_____ Heat energy from the sun causes the particles of liquid water to move faster and faster.

_____ The liquid water evaporates, or becomes a gas.

_____ The particles go into the air.

3. Draw a picture of something wet. Then draw what it looks like after the water evaporates.

✓Concept Check

1. A **Cause** is something that makes another thing happen. An **Effect** is the thing that happens. Look for a cause and effect on these two pages. Underline the cause and circle the effect.

2. Circle the word that means a change from a liquid to a gas.

   melt **evaporate** **freeze**

3. When water boils, what do the bubbles show?

____

Boiling and Freezing

Boiling is also a change from a liquid to a gas. When we boil a liquid, we add heat energy until it bubbles. The bubbles are the gas form of the liquid. Boiling works faster than evaporation does.

Heat energy causes the water to boil. The bubbles show that the liquid water is becoming a gas. ▶

◀ This solid is called dry ice. It changes directly from a solid to a gas. Solid water (ice) first melts to become a liquid. Then it boils or evaporates to become a gas.

Freezing is the change from a liquid to a solid. When you put liquid water in a freezer, the cold air slows the particles. They lose energy. When the water particles get slow enough, the water freezes. It becomes ice.

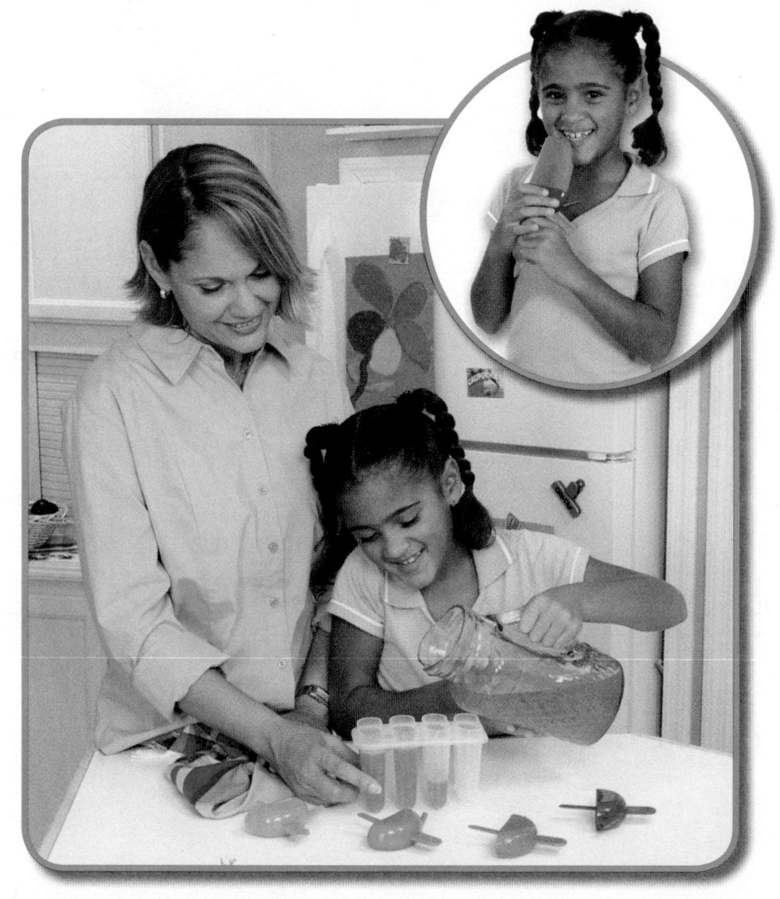

▲ **The girl and her mother pour juice into molds. They put the molds in a freezer. The juice freezes into ice pops.**

✓Concept Check

1. What causes water to boil and to freeze?

Boil:____________ Freeze: ____________

2. Circle the word that is the opposite of melting.

boiling **evaporation** **freezing**

3. When you put liquid water in the freezer, what does the cold air do to the particles?

____

4. What happens when the water particles get slow enough?

____

5. What is another word for frozen water?

____

✓Concept Check

1. A **Cause** is something that makes another thing happen. An **Effect** is the thing that happens. Look for a cause and effect on these two pages. Underline a cause and circle its effect.

2. Why does water form on the outside of the glass?

____

____

3. Use the terms to finish the chart. Match the action with the state of matter it creates.

condensation **melting** **boiling**
freezing **evaporation**

Solid	Liquid	Gas

Condensing

There is water on the outside of the glass in the picture. Is the glass leaking? No. The air around the glass has water vapor in it. *Water vapor* is the gas form of water.

The particles of water vapor next to the cold glass lose energy. They turn back into drops of liquid water on the outside of the glass. The water vapor *condenses*. Condensing is the opposite of evaporating.

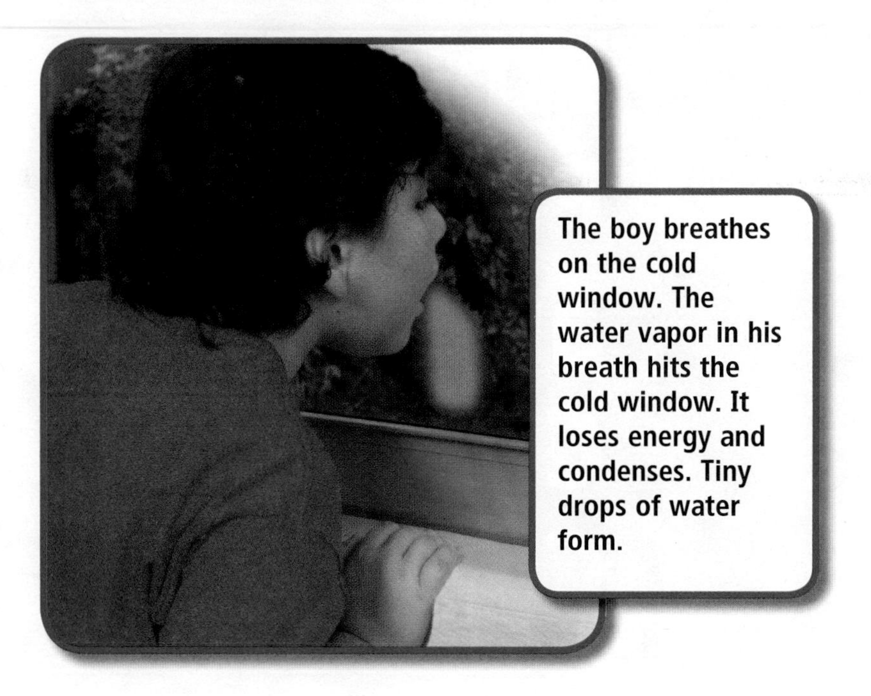

The boy breathes on the cold window. The water vapor in his breath hits the cold window. It loses energy and condenses. Tiny drops of water form.

Lesson Review

Fill in these Cause and Effect sentences.

1. Matter can change form when it is made __________ or __________.
2. If you put an ice cube in a pot of boiling water, it will __________.
3. If you put a wet towel in the sun, the water in the towel will __________.
4. If you put juice in the freezer, it will __________.

California Standards in This Lesson

 **I.g** *Students know* that when two or more substances are combined, a new substance may be formed with properties that are different from those of the original materials.

Vocabulary Activity

A noun is a person, place, or thing. A verb is an action word.

1. Is the word mixture a noun?
   How do you know?

   ___

2. There is a verb in the word mixture.
   What is it?

   ___

3. Write a sentence using the word mixture and the verb mix.

   ___

VOCABULARY
mixture

What Are Physical and Chemical Changes?

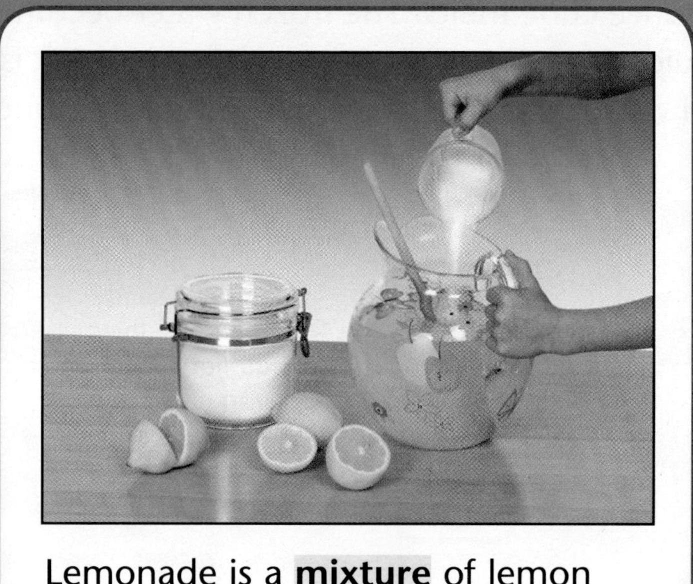

Lemonade is a **mixture** of lemon juice, water, and sugar.

Hands-On Activity
Changes

1. List three chemical changes and three physical changes that you have seen or know about. When you have named three of each, share the list with a friend.

Chemical Changes	Physical Changes
____________________	____________________
____________________	____________________
____________________	____________________

2. Fruit salad is a kind of mixture. Ask an adult to help you make a fruit salad. Write down the kinds of fruit you put in your mixture.

____________________ ____________________

____________________ ____________________

____________________ ____________________

✓Concept Check

1. The **Main Idea** on these two pages is Physical changes happen when no new substance is formed. **Details** tell more about the main idea. Find details about how physical changes happen when no new substance is formed. Underline two of them.

2. What kinds of physical changes can you make to a pencil?

____

3. Circle the word or words that make the statement true.

   1. Changes of state are (physical chemical) changes.

   2. If you mix salt and sand together, they (do do not)change into new matter.

   3. A (chemical change physical change) happens when no new substances are formed.

Physical Changes

Glass breaks. It is still glass. You tear a piece of paper. It is still paper. These are physical changes. A *physical change* happens when no new substances are formed.

An ice cube melts. The frozen water becomes liquid water. The water changes state, but it is still water. Changes of state are physical changes.

When you mix cereal, bananas, and strawberries in a bowl, you get a mixture. A **mixture** is a *combination* of two or more kinds of matter.

Mixing is a physical change. The cereal and the fruit have not become different substances. You could pick the fruit out of the cereal if you wanted to.

If you put drink mix into water and stir, it will disappear. The drink mix *dissolves.* If you let the water evaporate, the drink mix will be left behind as a solid. *Dissolving* is a physical change.

✓Concept Check

1. Give one example of a physical change.

____

____

____

2. Another word for mixture is ____________.

3. Draw a picture showing another mixture that you eat.

✓Concept Check

1. The **Main Idea** on these two pages is In a chemical change, substances react to form new kinds of matter. **Details** tell more about the main idea. Find details about how substances react to form new kinds of matter. Underline two of them.

2. Tell whether each change is chemical or physical.

Change	Chemical or Physical
Wood is burned.	
Ice is melted.	
Cream is mixed with coffee.	
A cut apple turns brown.	

3. Can the brown substance on the apple change back to apple? Why or why not?

____

Chemical Changes

If you cut an apple and leave it out in the air, the apple turns brown. This is a chemical change.

In a *chemical change*, substances react to form new kinds of matter. The oxygen in the air reacts with the apple. The brown substance is a new kind of matter.

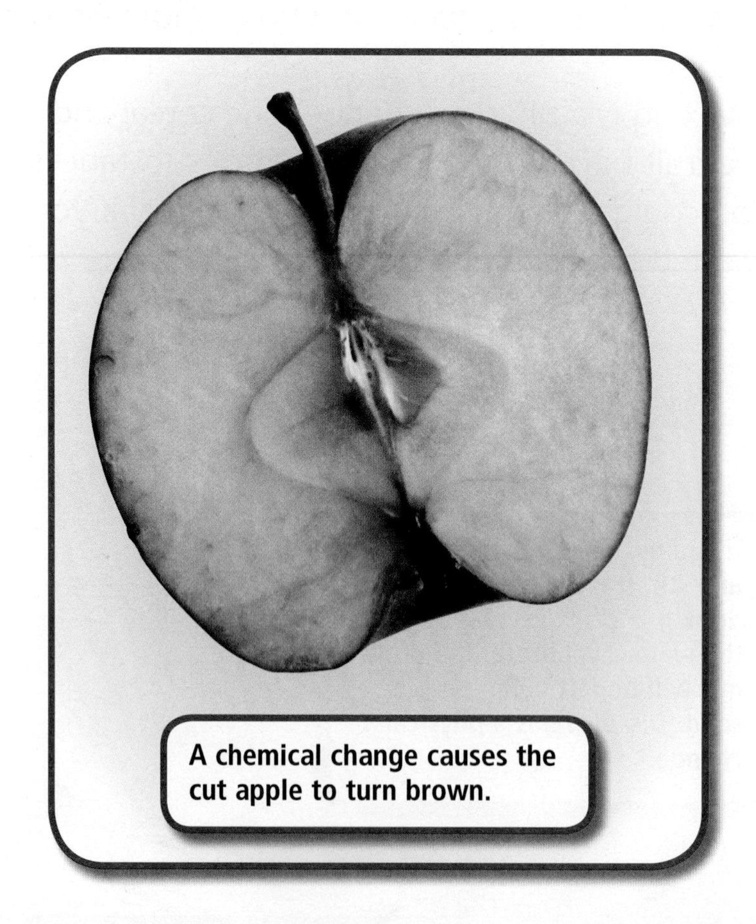

A chemical change causes the cut apple to turn brown.

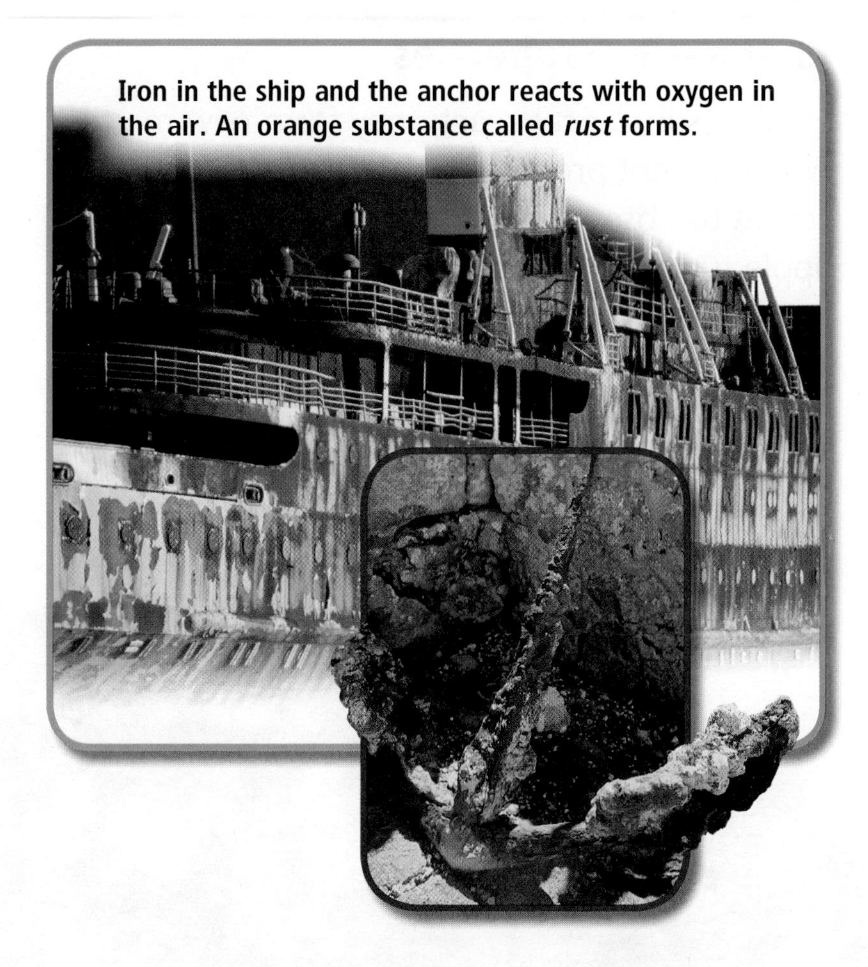

Iron in the ship and the anchor reacts with oxygen in the air. An orange substance called *rust* forms.

The ship and the anchor shown above have rusted. *Rusting* is a chemical change. It happens when oxygen in the air reacts with iron. Rust is different from both iron and oxygen. Rust cannot be changed back into iron.

✓Concept Check

1. How do you know that rusting is a chemical change?

____

2. Chemical changes are sometimes written as formulas. Finish this formula for rust.

__________ + Oxygen = Rust

3. The ship's owners remove the rust from the anchor by scraping it off. Is this a chemical change or a physical change?

____

✓Concept Check

1. The **Main Idea** on these two pages is You can use your senses to observe chemical changes. **Details** tell more about the main idea. Find details about how physical changes happen when matter changes but no new substance is formed. Underline two of them.

2. Write two signs of each chemical change.

Fireworks: ______________________________

____

Wood burning: ______________________________

____

Iron rusting: ______________________________

____

Changes in Properties

When substances react, they form new matter with different properties. You can use your senses to observe these changes. You can see the apple turn brown. You can see the orange color of rust. As wood burns, you can smell and see smoke. You can see the wood turn to ash.

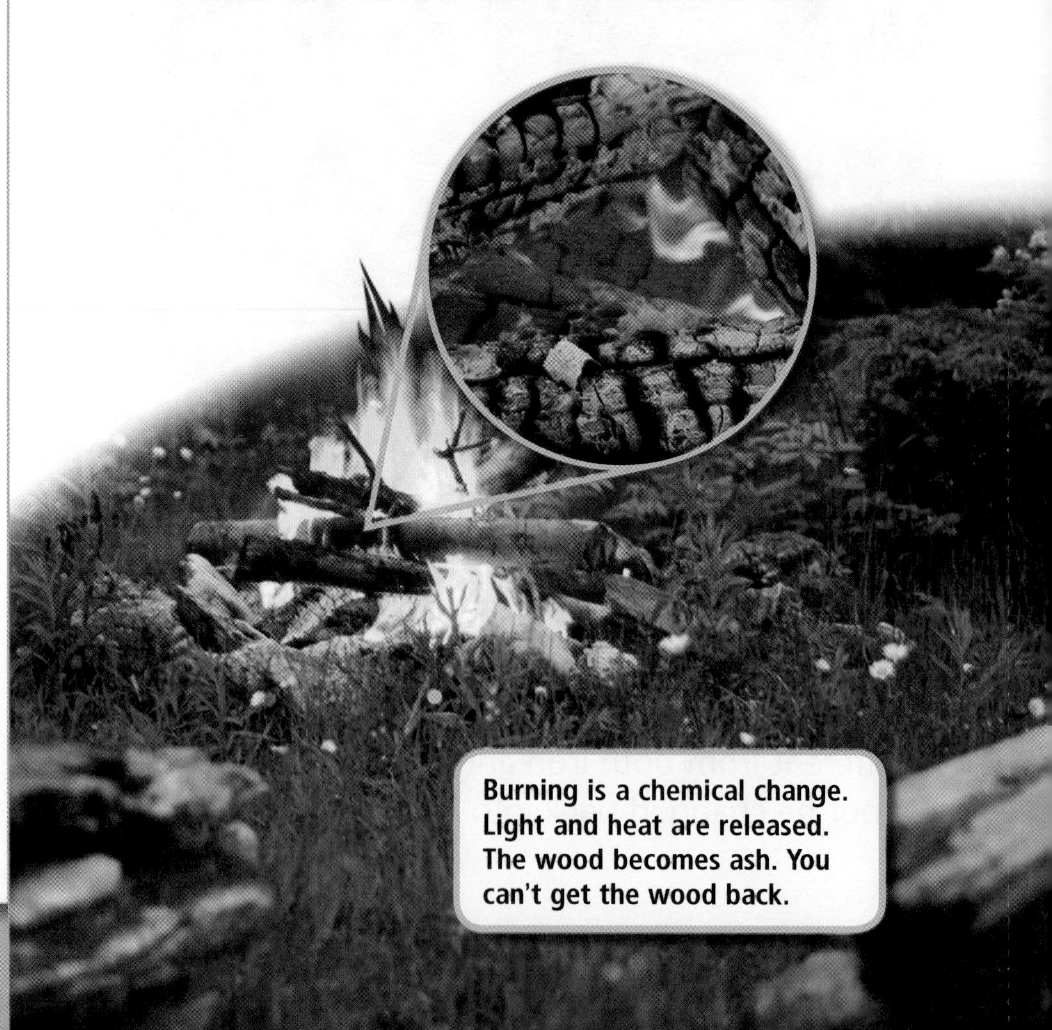

Burning is a chemical change. Light and heat are released. The wood becomes ash. You can't get the wood back.

Have you ever watched fireworks? You can see the sky light up with color. You can hear the loud pops and booms. These things are signs that chemical changes are happening.

As the fireworks go off, matter changes. You can't get the fireworks back.

Lesson Review

Complete this Main Idea sentence.

1. ____________ can go through physical and chemical changes.

Complete these Detail statements.

2. A ____________ change happens when no new substance or kind of matter forms.

3. A ____________ change happens when substances react to form new matter.

4. You can use your senses to observe how ____________ of matter change.

California Standards in This Lesson

 **I.h** *Students know* all matter is made of small particles called atoms, too small to see with the naked eye.

 **I.i** *Students know* people once thought that earth, wind, fire, and water were the basic elements that made up all matter. Science experiments show that there are more than 100 different types of atoms, which are presented on the periodic table of the elements.

Vocabulary Activity

atoms

element

Use the vocabulary words above to answer the following question.

1. Are all elements made of atoms?

____

2. Is every kind of matter an element?

____

____

Lesson 7

VOCABULARY
atom
element

What Are Atoms and Elements?

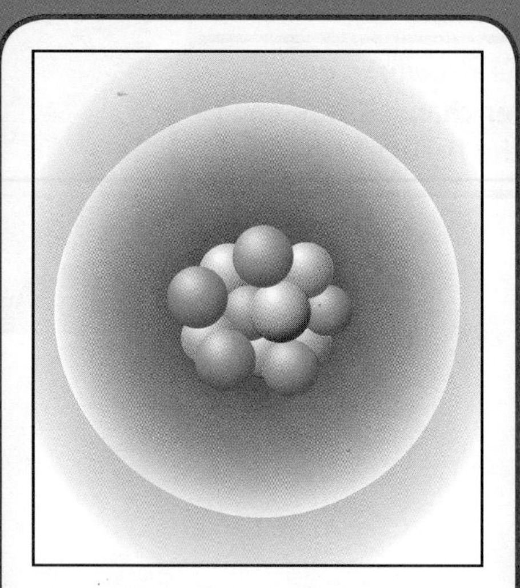

All matter is made of tiny particles called **atoms**. An atom is too small to see with the eye alone.

This wire is made of copper. Copper is an **element**. An element has only one kind of atom.

Hands-On Activity
Elements

1. Look at this part of the Periodic Table of the Elements.

15	16	17	18
			2 He Helium
7 N Nitrogen	8 O Oxygen	9 F Fluorine	10 Ne Neon
15 P Phosphorus	16 S Sulfur	17 Cl Chlorine	18 Ar Argon
33 As Arsenic	34 Se Selenium	35 Br Bromine	36 Kr Krypton

Find one element that is not used in the lesson. What symbol did you find? What is the full name of the element?

____

2. Ask an adult to help you find something in your house that is an element. Draw what you found. Then write the name of the element.

✓Concept Check

1. The **Main Idea** on this page is All matter is made of atoms. **Details** tell more about the main idea. Find details about atoms. Underline two of them.

2. What is an atom?

____

____

____

3. Order the matter below from smallest to largest using the numbers 1, 2, and 3, with 1 being the smallest.

_____ grain of sand

_____ atom

_____ sand castle

Atoms

Have you ever built a sand castle? It looks like a solid building. Up close, you can see that it is made up of many tiny grains of sand.

All matter is made up of tiny particles called atoms. An **atom** is the smallest particle of matter that has the properties of that matter. Atoms are much, much smaller than grains of sand.

A grain of sand is many, many times larger than an atom.

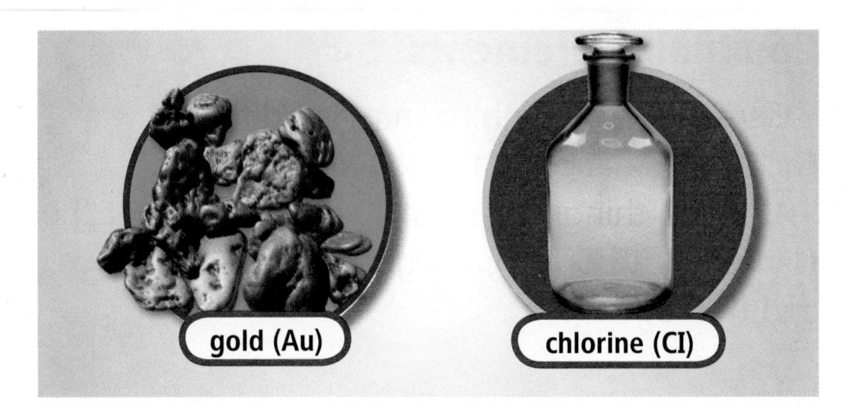

Elements

All the atoms in a piece of gold are alike. All the atoms in chlorine gas are alike. Gold and chlorine are elements. An **element** has only one kind of atom.

More than 100 elements are shown in a chart called the *periodic table*. It gives each element's name. It also gives each element's *symbol*. This is one or two letters that stand for the element.

✓Concept Check

1. What is an element?

____

____

2. What is the chart of elements called?

____

3. Match the name of the element with its symbol.

Element	Symbol
gold	Cl
chlorine	Hg
mercury	Ag
silver	Au

✓Concept Check

1. The **Main Idea** on these two pages is Each combination of elements is a different substance. **Details** tell more about the main idea. Find details about combinations of elements. Underline two of them.

2. How many elements make up most living things? Circle your answer.

   2 3 4 5

3. Water is made of the elements ____________ and ____________.

4. One particle of water looks like this:

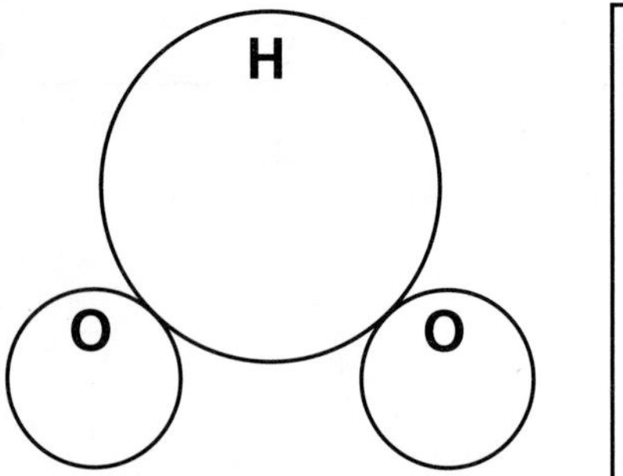

Draw what a second water particle would look like.

Combining Elements

Elements can combine in many different ways. Each combination is a different substance.

All living things are mostly combinations of the elements carbon (C), oxygen (O), nitrogen (N), and hydrogen (H).

Every grain of sand is made of many tiny particles. Each particle has two kinds of atoms.

Water is not listed on the periodic table. Why not? Water is not an element. It is made of two kinds of elements. These elements are hydrogen (H) and oxygen (O).

Every particle of water contains the same two elements.

Lesson Review

Complete this Main Idea statement.

1. An __________ is the smallest particle of matter that has all the properties of that matter.

Complete these Detail statements.

2. An __________ is a substance that has only one kind of atom.
3. More than 100 elements are shown in the __________.
4. Elements can __________ to make many different substances.

Unit 1 Review

Circle the letter in front of the best choice.

1. Every _______ needs energy.

   A ball

   B action

   C atom

   D electricity

2. What types of energy come from the sun?

   A heat and light

   B heat and wind

   C wind and electricity

   D water and light

3. What turns the sun's energy into food?

   A wind farm

   B animal

   C plant

   D battery

4. Fuel is used to move or _______ things.

   A heat

   B shrink

   C break

   D open

5. Light and heat waves are _______.

   A visible

   B large

   C strong

   D invisible

6. Everything around us is _______.

   A energy

   B matter

   C waves

   D solids

7. Volume is the amount of _______ an object has.

   A mass

   B length

   C space

   D energy

8. Which is not an example of a solid?

   A spoon

   B blanket

   C hair

   D lemonade

9. Heating a liquid until it becomes a gas is called ________.

   A freezing

   B condensing

   C boiling

   D melting

10. Matter that goes through a chemical change cannot be ________.

   A burned

   B changed back

   C frozen

   D broken apart

11. Explain how energy comes from the wind and becomes electricity that lights a lamp in your classroom.

____

____

____

____

____

____

12. Explain why it would be safer to be outside during an earthquake.

____

____

____

13. Name the states of matter. Give an example of each.

____

____

14. Look back to the question you wrote in the Study Journal. Do you have an answer for your question? Tell what you learned that helps you understand energy and matter.

____

____

____

____

____

Light

In this unit, you will learn about how light travels, what causes color, and how you see objects. What do you know about these topics? What questions do you have?

Thinking Ahead

How does light travel from a lamp to a book? Add to the picture to show how light travels.

What causes color?
Write two of your ideas.

____

____

____

How are you able to see the things around you? Write what you think.

____

____

Write a question you have about light.

____

____

Recording What You Learn

On this page, record what you learn as you read the unit.

Lesson 1

Draw two pictures that show details about how light travels.

Picture 1	Picture 2

Lesson 2

What causes the colors you see? Write a fact about how colors appear.

____

____

____

Lesson 3

How are you able to see objects? Fill in the boxes to show the sequence.

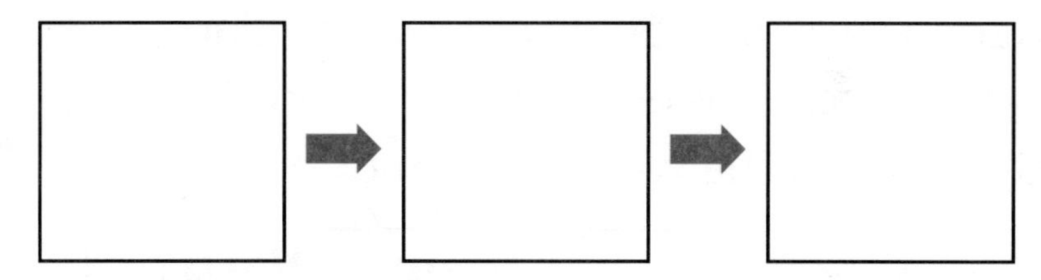

How does the eye help you see an object? Write what you think.

____

____

____

California Standards in This Lesson

 **2.a** *Students know* sunlight can be blocked to create shadows.

 **2.b** *Students know* light is reflected from mirrors and other surfaces.

Vocabulary Activity

1. Read the definition of each vocabulary word. Decide how each word is used in a sentence. Is the word a noun or a verb?

Word	Noun or Verb?
shadow	
reflect	

2. Look at pages 58 and 59. Underline each sentence that has a vocabulary word in it. Then draw an arrow from the sentence to what it tells about in the picture.

Lesson 1

How Does Light Travel?

VOCABULARY
shadow
reflect

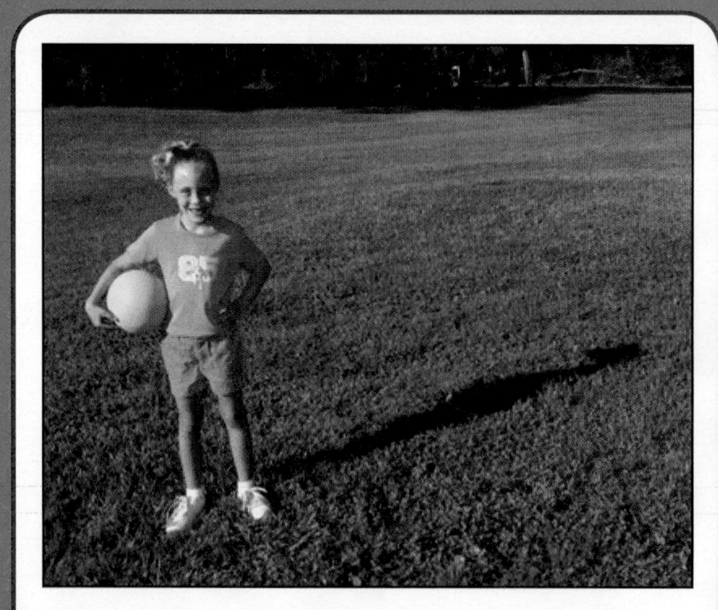

The girl's body blocks light coming from the sun. A **shadow** appears on the ground.

The girl looks in a mirror. The mirror **reflects** her image. She sees the image of her body in the mirror.

Hands-On Activity
Observe

Some calculators use a battery to get the energy to run, but solar calculators do not. They use a solar cell, which takes in light.

1. Get a solar calculator. Turn on the calculator. Record what happens.

____

____

2. Find the solar cell. Cover the solar cell with your finger. What happens?

____

3. Take your finger away. What happens?

____

How do you know that light is a form of energy?

____

____

✓Concept Check

1. The **Main Idea** on these two pages is Light is a form of energy. **Details** tell more about the main idea. Find details about light. Underline two of them.

2. Underline the sentence that tells the main idea.

3. What changes the energy of sunlight into food?

____

4. How do lights help the drivers of dune buggies?

____

5. Circle something in the picture that shows light is energy you can see.

Light Energy

Light is a form of energy. You can see light energy. Light comes from different things. It can come from fire, light bulbs, and the sun.

Plants change the energy of sunlight into food that helps them grow.

The lights on these dune buggies help the drivers see at night.

People use light in many ways. A light bulb helps people see in the dark. Traffic lights help people drive safely. People can eat apples that used light energy to grow.

✓Concept Check

1. Circle the sentence on page 60 that tells three examples of where light comes from.

2. Fill in the chart to show how people use light.

How People Use Light
A.________________
B.________________
C.________________

3. How does light get from the lamp to the girl's book? Draw an arrow that shows how the light travels.

✓Concept Check

1. The **Main Idea** on these two pages is Light travels. **Details** tell more about the main idea. Find details about how light travels. Underline two of them.

2. How does a shadow form?

____

____

3. Why can you see your face in a mirror?

____

____

____

____

4. Look at the picture on page 62. What would happen to the shadow if you moved the light?

____

Shadows

The light from a flashlight shines out in straight lines. The sun shines down in straight lines. When something stops or blocks light, a dark area forms. This dark area is a **shadow**.

The vase of flowers blocks the light from the lamp. A shadow forms behind the vase.

Bouncing Light

When you look in a mirror, you see light that bounces off your body. The light moves in a straight line and hits the mirror. The light **reflects**, or bounces, off the mirror. You can see your *reflection* in the mirror. You can see reflections in mirrors and in smooth water.

The trees and the mountain are reflected in the lake.

Lesson Review

Complete this Main Idea sentence.

1. __________ can travel.

Complete these Detail sentences.

2. Light is a form of __________.
3. A dark area called a __________ forms where an object blocks light.
4. Light __________, or bounces, off smooth water.

California Standards in This Lesson

 **2.c** *Students know* the color of light striking an object affects the way the object is seen.

Vocabulary Activity

1. Circle words for things that can absorb something.
2. Underline words for things that are absorbed.

Lesson 2

VOCABULARY
absorb

What Causes Color?

A sponge can **absorb**, or take in, water. The color black takes in light. A black street absorbs all the colors of light.

Hands-On Activity
Experiment, Draw Conclusions

1. Find a bowl, a small container of water, and three different kinds of paper.
2. Place one piece of paper over the bowl. Put a drop of water on the paper. Do the same thing with each kind of paper.
3. What happens to the water and the paper?

____

____

4. Do all kinds of paper absorb the same amount of water? Why, or why not?

____

____

____

The same is true of objects and light. Some objects absorb more light energy than others do.

✓Concept Check

1. A **Cause** is something that makes another thing happen. An **Effect** is the thing that happens. Underline a cause, and circle its effect.

2. Find two more examples of a cause and its effect. Finish this graphic organizer.

The Cause	The Effect
Rain falls.	
Light bends.	

3. What happens when white light is made to bend?

____

____

Light and Color

Do you know how a rainbow forms? Rain and sunlight are needed. After rain falls, little drops of water stay in the air. Light from the sun shines through the drops of water. The raindrops bend the sunlight. The bending of light makes the colors spread out, or separate.

The separated colors form a rainbow. You can see red, orange, yellow, green, blue, and violet light.

◀ **Light bends as it passes through drops of water in the air. The light separates into colors. We see a rainbow.**

You can change light with a prism. A *prism* is a special piece of glass. Light bends as it passes through a prism. The prism causes white light to separate into colors.

✓Concept Check

1. Look at the picture. What caused white light to split into different colors?

___

2. Draw a line under the sentence that tells what a prism is.

3. Fill in this graphic organizer to show what a prism does.

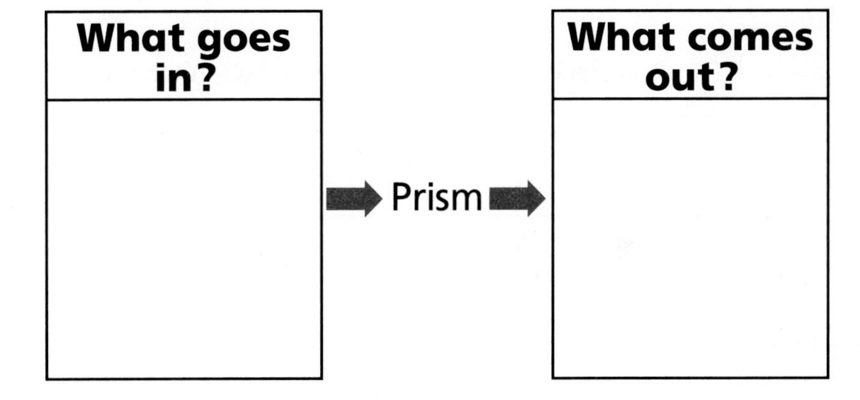

4. Underline the sentence that tells the colors we see in a rainbow.

✓Concept Check

1. A **Cause** is something that makes another thing happen. An **Effect** is the thing that happens. Look at the caption on the picture of the prism. Underline a cause, and circle its effect.

2. What colors are needed to make white light?

____

3. What is needed to show the true color of something?

____

4. Circle the prism shown on this page.

Objects in White Light

You have learned that white light can separate into different colors.

All the colors together make white light. The sun and a clear light bulb shine with white light. White light shows the true colors of things.

White light enters the prism. Different colors of light come out.

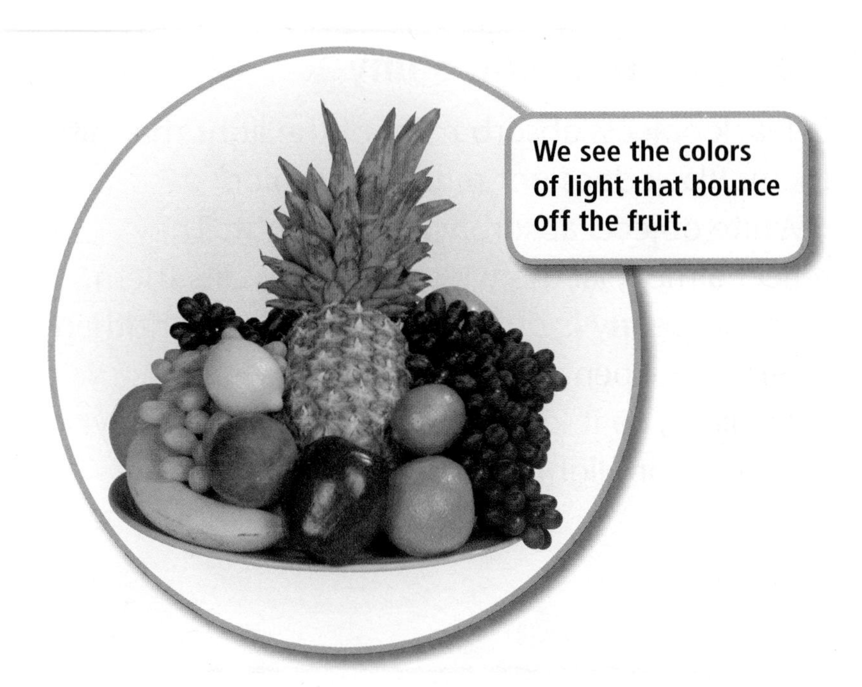

When white light hits an object, the object **absorbs**, or takes in, some of the colors in the light. Other colors of light bounce off the object.

When white light hits a lemon, yellow light bounces off. We see the yellow light that is reflected from the lemon. A lemon absorbs all the other colors of light. This is why a lemon looks yellow.

Green light reflects off a lime. A lime absorbs all colors of light except green. We see the green lime.

✓Concept Check

1. What causes a lemon to look yellow?

____

____

____

2. Why does the apple look red? Finish this graphic organizer.

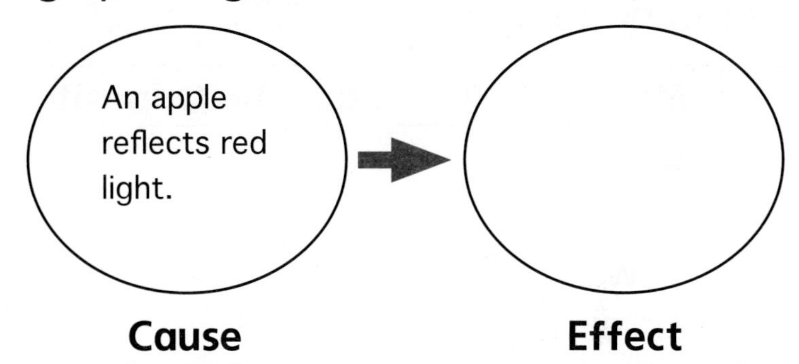

3. Underline the words that mean the same thing as *reflect.*

4. Circle a word that means the opposite of *reflects.*

✓Concept Check

1. A **Cause** is something that makes another thing happen. An **Effect** is the thing that happens. Two sentences on this page have a cause and an effect. Underline one sentence, and circle the other.

2. Some words on this page are opposites. Fill in this graphic organizer to show the opposites.

Word	Means the Opposite
absorb dull new darker	

3. Underline the sentence that tells why a lemon looks black in red light.

Black, White, Dull, Shiny

Black objects absorb most of the light that hits them. Black does not reflect any color.

White objects absorb very little light. They reflect almost all the white light that hits them.

An old penny is dull. It absorbs more light than a shiny new penny. The shiny penny reflects more light, so it looks brighter. The dull penny absorbs more light, so it looks darker.

Objects absorb some colors of light and reflect others.

Objects in Colored Light

In white light, a lemon looks yellow. If you shine a red light on a lemon, it will look black. The lemon absorbs all the red light. There is no yellow light for the lemon to reflect.

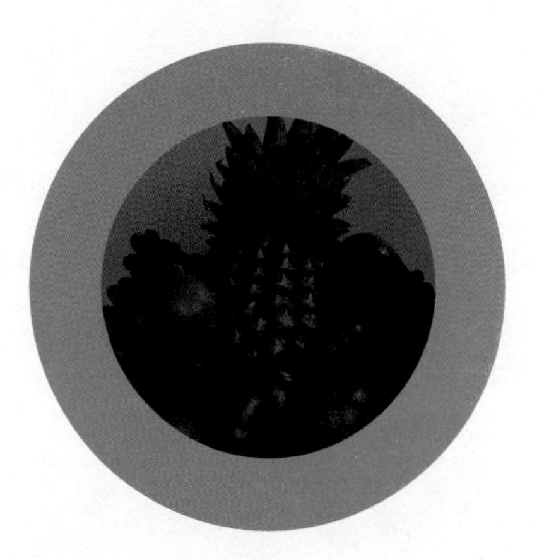

Lesson Review

Complete these Cause and Effect sentences.

1. Bent sunlight causes a __________ to form.
2. A piece of glass called a __________ causes light to separate into colors.
3. Because a lime __________ the color green, it looks green.
4. Because a black street __________ most of the light that hits it, it looks black.

California Standards in This Lesson

 **2.a** *Students know* an object is seen when light traveling from the object enters the eye.

Vocabulary Activity

The word opaque comes from a Latin word that means "shady." This word is similar to a word you learned in Lesson 1.

1. Turn back to page 62 in Lesson 1. What word on this page is similar to the word *shady?* Tell what this word means.

____

____

2. Look at the picture of the boy holding a plate. What is another thing that is *opaque?*

____

How Do You See Objects?

VOCABULARY
opaque

The plate is opaque. You cannot see through **opaque** objects. You cannot see all of the boy's face.

Hands-On Activity
Observe

1. Put a white cotton ball on your desk. Look at the cotton ball through a piece of red cellophane. How does the cotton ball look?

____

____

2. Is the cellophane opaque? Explain your answer.

____

____

3. Does the cellophane block any of the light that comes from the cotton ball? Explain your answer.

____

____

✓Concept Check

1. To put events in **Sequence** is to put them in the order in which they happen. When our eyes see objects, a sequence of events happens. A paragraph on this page tells about a sequence of events. Circle the paragraph.

2. Light bounces off of objects. What happens next for you to see the object?

____

3. Underline two sentences that tell how light travels.

4. In what sequence does light travel when you see a tree? Fill in the boxes. Use the words **eyes** and **tree**.

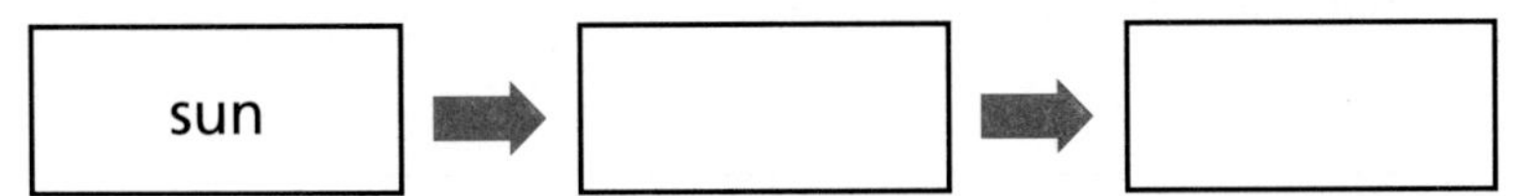

Seeing Light

Light is all around you. The sun gives off light. Lamps and fires give off light. Light travels in straight lines. Light bounces off objects.

In a forest, light bounces off the trees. Then the light travels to your eyes. When the light reaches your eyes, you see the trees.

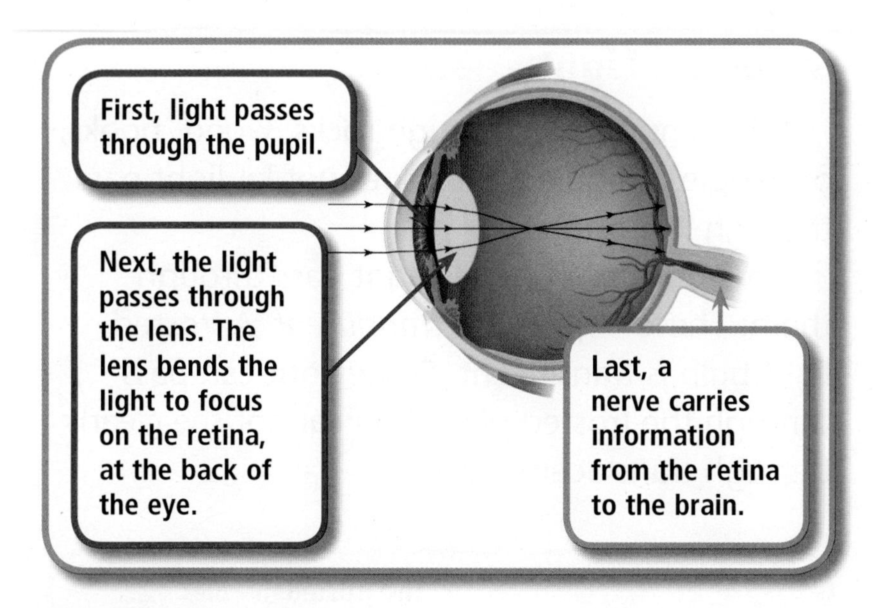

Light and the Eye

Light enters the eye through an opening called the *pupil*. The colored part around the pupil is the *iris*. In bright light, the iris makes the pupil small. Only a little light can get through. In the dark, the iris makes the pupil large. More light can get through.

The light then passes through the lens. The *lens* bends the light. The light shines on the retina. The *retina* curves around the back of the eye. A nerve carries information about the light to the brain. Your brain tells you what you see.

✓Concept Check

1. Does light first enter the pupil of your eye or the lens?

____

2. In what sequence does light travel when it enters the eye? Use the words **lens**, **retina**, and **pupil.**

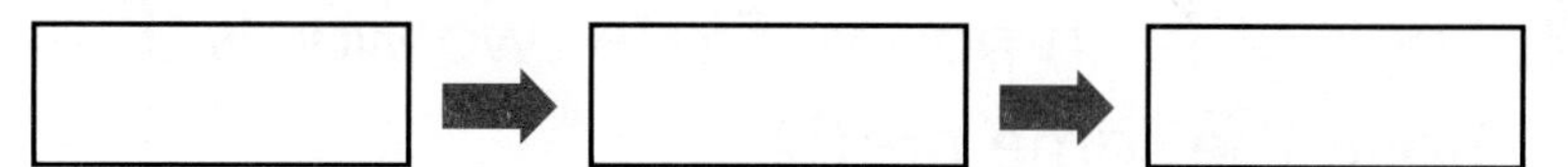

3. What part of the eye makes the pupil small or large?

____

4. What part of your body tells you what you see?

____

✓Concept Check

1. To put events in **Sequence** is to put them in the order in which they happen. When our eyes see objects, a sequence of events happens. Light hits a translucent object. What happens next?

____

2. Look at both pages. Circle two words that have the same prefix.

3. Look at the pictures on these pages. Then complete this graphic organizer. Use the words *opaque*, *translucent*, and *transparent*.

Kind of Material	Effect on Light
clear glass	
stones	
frosted marbles	

Stopping Light

Light cannot pass through rocks, wood, books, or people. An object that does not let light pass through it is **opaque**.

Some objects let a little light pass through them. These objects are *translucent*. A frosted light bulb is translucent. Some light can pass through the frosted glass. You cannot see clearly through translucent glass.

The frosted marbles are translucent. You can see some light through them. ▼

▲ The stones are opaque. You cannot see through them.

A clear window or a clear vase is *transparent*. It lets most of the light pass through. You can see clearly through transparent glass.

The clear vase is transparent. ▶

Lesson Review

Complete these Sequence sentences.

1. You see an object when light bounces off it and travels to your __________.
2. Light first enters the eye through the __________.
3. A nerve carries the information from the retina to the __________.
4. When light reaches a __________ object, some of it passes through.

Circle the letter in front of the best choice.

1. Which word means "to bounce off"?

   A absorb

   B opaque

   C reflect

   D shadow

2. Which sentence tells how a shadow forms?

   A Light passes through an object.

   B Light is blocked by an object.

   C Light goes around an object.

   D Light is changed by an object.

3. On a walk, you see your shadow. Where does the light that causes your shadow come from?

   A a cloud

   B a tree

   C the ground

   D the sun

4. Which is one way people use light?

   A to see in the dark

   B to give a car energy

   C to make our own food

   D to get energy to grow taller

5. Which object reflects light like a mirror?

   A a red apple

   B a smooth lake

   C a rough ocean

   D a dark shadow

6. How does light travel from place to place?

   A in straight lines

   B in curved lines

   C in many circles

   D in large circles

7. What piece of glass bends light?

   A mirror

   B prism

   C vase

   D color

8. What causes sunlight to make a rainbow?

____

____

9. Which color of light must shine on an apple for it to look red?

   A black

   B white

   C blue

   D green

10. Which part of the eye changes so more or less light can enter?

   A nerve

   B lens

   C iris

   D retina

11. Which reflects the most light?

   A a shiny penny

   B a dull penny

   C a black penny

   D a red penny

12. Which of these objects is opaque?

   A clear glass

   B the air

   C pure water

   D your body

13. Why does a lemon look yellow?

____

____

14. You see a flower. Which shows the sequence in which light travels?

   A flower, sun, eye

   B sun, flower, eye

   C eye, flower, sun

   D sun, eye, flower

15. Look back at the question you wrote on page 56. Do you have an answer for your question? Tell what you learned that helps you understand how light travels.

____

____

____

____

Adaptations

In this unit, you will learn about how plant parts help plants, how animal parts help animals, and what lives in different environments. You will also learn about changes in environments, and how living things have changed over time. What do you know about these topics? What questions do you have?

Thinking Ahead

How do the parts of living things help them stay alive? Draw an arrow to a part that helps each living thing below stay alive.

How do living things change environments?

____

____

What is one way changes to environments affect living things?

____

____

How have living things changed over time?

____

____

Write a question you have about the adaptations living things make to survive in their environments.

____

____

Recording What You Learn

On this page, record what you learn as you read the unit.

Lessons 1 and 2

What adaptations do living things have that help them live in their environments? Draw a picture of a plant with an adaptation and a picture of an animal with an adaptation.

Plant	Animal

Lesson 3

What living things live in different environments? Write the name of the environment in which each living thing you drew lives.

Plant: ______________________

Animal: ______________________

Lessons 4 and 5

How do living things change environments? Give one example of a harmful change and one example of a helpful change.

____

____

Lesson 6

How do living things change over time? Suggest a sequence of events that might happen by completing this graphic organizer.

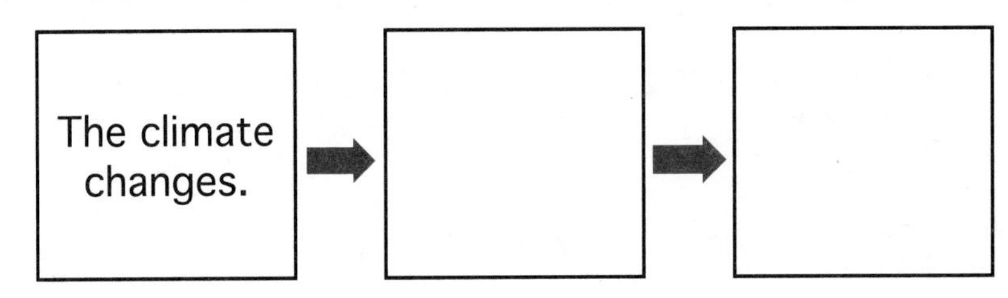

California Standards in This Lesson

 **3.a** *Students know* plants and animals have structures that serve different functions in growth, survival, and reproduction.

 **3.b** *Students know* examples of diverse life forms in different environments, such as oceans, deserts, tundra, forests, grasslands, and wetlands.

Vocabulary Activity

trait **adaptation**

survive **reproduce**

Use the vocabulary words above to complete the following sentences.

1. When living things ____________ they have young.
2. The name for a body part that helps a plant or animal __________ is __________.
3. An adaptation is a special kind of ____________.

Lesson 1

How Do Plant Parts Help Plants?

VOCABULARY
trait
survive
adaptation
reproduce

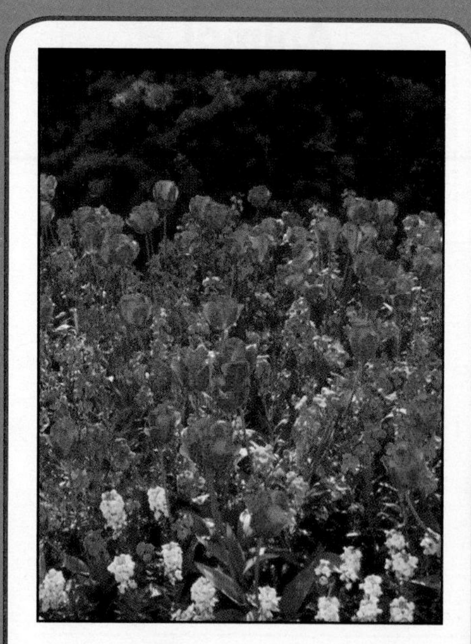

One **trait** of a flower is its color.

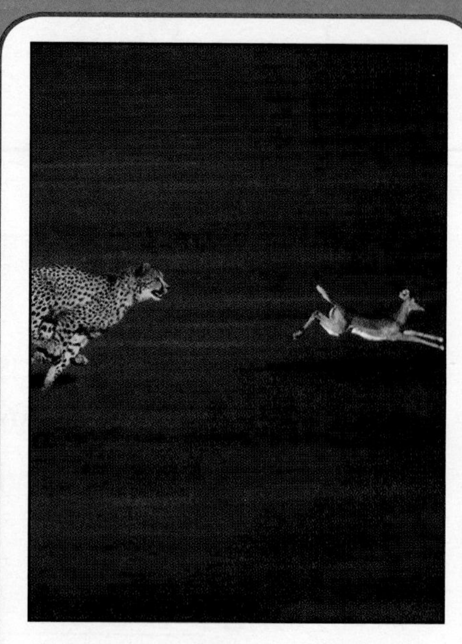

Running fast helps animals **survive**, or stay alive.

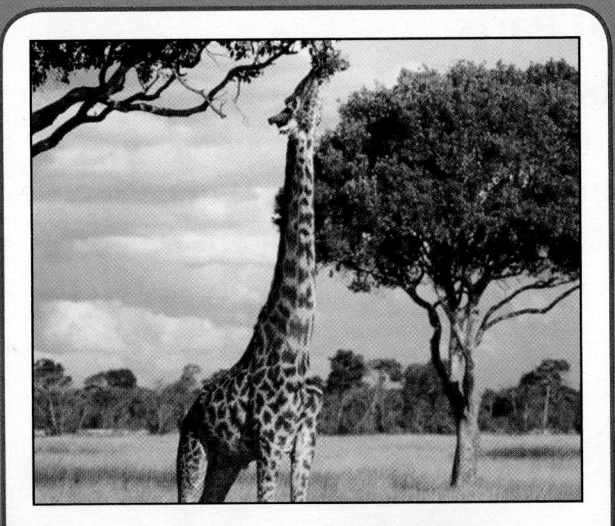

A long neck is an **adaptation** that helps a giraffe reach food.

When pigs **reproduce**, their young are called piglets.

Hands-On Activity
Observe and Compare

Look at two different kinds of plants. Find three traits for each plant. Write the traits for each plant below.

1. Traits of Plant 1:

___

___

___

2. Traits of Plant 2:

___

___

___

3. How are the two plants different?

___

___

✓Concept Check

1. The **Main Idea** on these two pages is Adaptations are traits that help a living thing survive. **Details** tell more about the main idea. Find details about how adaptations help plants survive. Underline two of them.

2. What is one adaptation of a rose?

____

3. In the picture, circle a trait of the Venus' flytrap.

4. What trait do all the plants shown on pages 84 and 85 have?

____

Plant Parts

A **trait** helps you tell things apart. Color, height, and shape are some traits.

Plants have traits that help them survive. Some plants have thorns. The thorns keep animals from eating the plants.

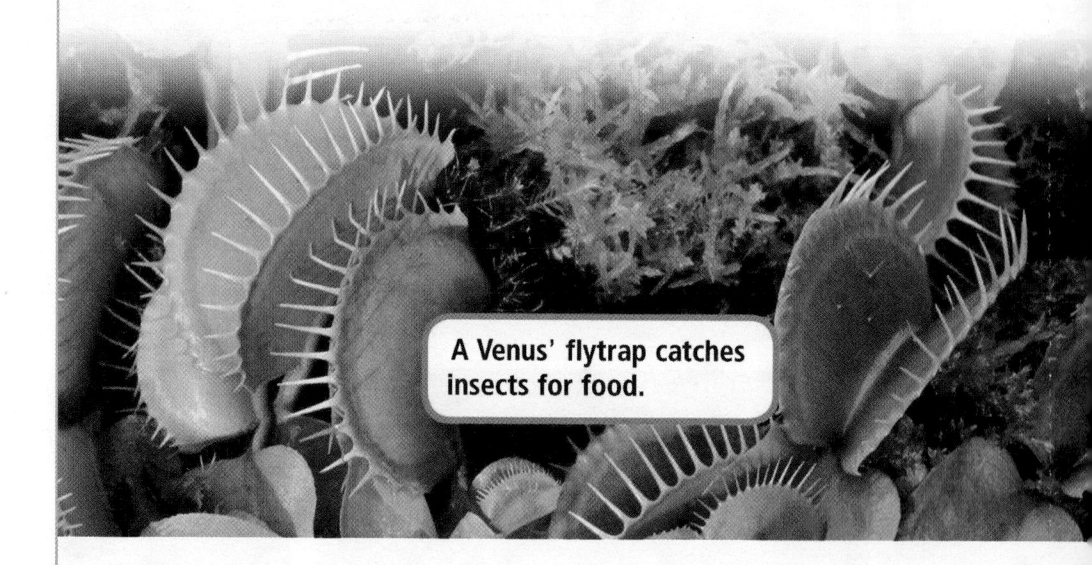

A Venus' flytrap catches insects for food.

These roots help the plant stand up.

Adaptations

Adaptations are special traits. They help a living thing **survive**, or stay alive. Some adaptations are parts. Wood is an adaptation. It is very strong. It helps trees grow tall without falling over.

Other adaptations are ways of acting. The way a plant acts can help it survive. A sunflower slowly turns all day to face the sun. This helps it get more sunlight to make food.

Thorns on a rosebush are an adaptation.

✓Concept Check

1. How does a sunflower get more sun?

____

2. What parts of plants are adaptations? Fill in the chart to show what you know.

Plant	Adaptation
Tree	
Rose	
Sunflower	

3. Circle an example of roots as an adaptation.

4. Do you think the trait you circled on page 84 is an adaptation? Explain your answer.

____

____

✓Concept Check

1. The **Main Idea** on these two pages is Adaptations help plants grow. **Details** tell more about the main idea. Find details about adaptations that help plants. Fill in this graphic organizer.

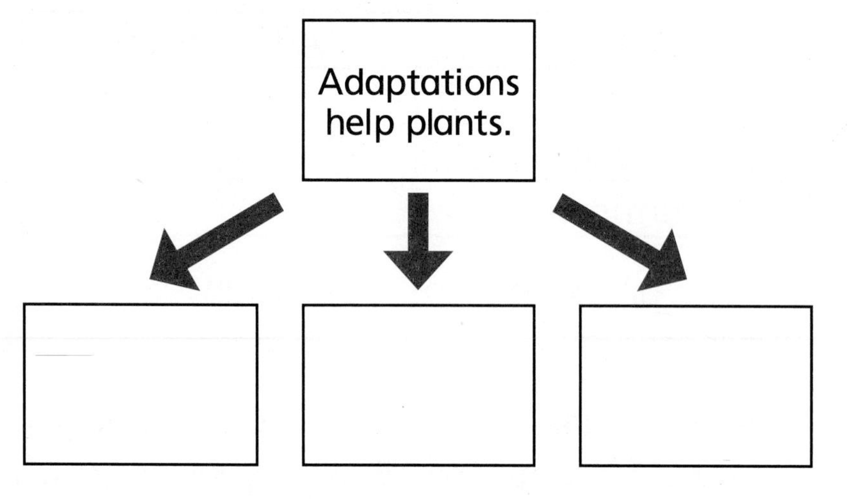

2. Tell how thick roots help trees survive.

____

____

Adaptations Help Plants

Adaptations help plants grow. Some plants have stems that can climb up trees. Then the leaves can get more sunlight to make food. This helps the plant grow better.

Some trees grow many thick roots. Then they can get more water to help the tree grow better.

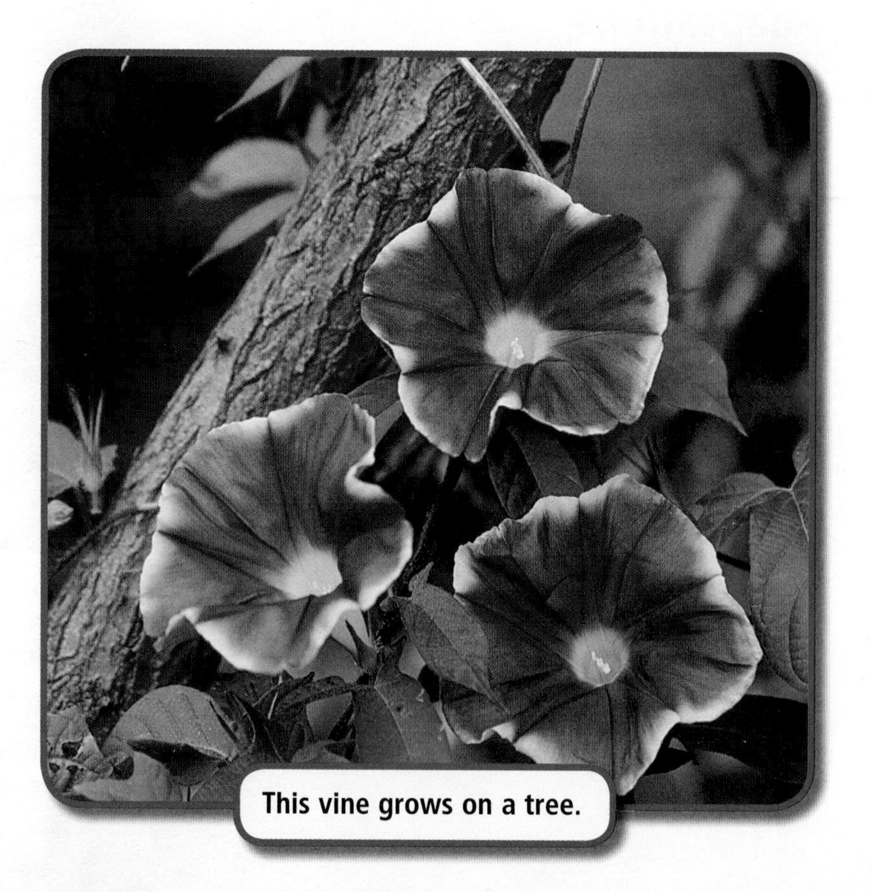

This vine grows on a tree.

Adaptations to Reproduce

All living things **reproduce**, or make young. Some plants use seeds to make new plants.

Flowers need pollen from other flowers to make seeds. They have bright colors and smells that attract bees. They have liquid that the bees sip for food. The bees carry pollen as they fly from one flower to another.

Bees help flowers reproduce. ▲

Lesson Review

Complete this Main Idea sentence.

1. An __________ is a trait that helps an animal or plant survive.

Complete these Detail sentences.

2. Some plants have __________ that can climb up trees.
3. Thick __________ help trees get more water to help them survive.
4. Bees help __________ reproduce by carrying pollen from flower to flower.

California Standards in This Lesson

 **3.a** *Students know* plants and animals have structures that serve different functions in growth, survival, and reproduction.

 **3.b** *Students know* examples of diverse life forms in different environments, such as oceans, deserts, tundra, forests, grasslands, and wetlands.

Vocabulary Activity

hibernate

migrate

Fill in the spaces with vocabulary words.

1. Animals that ___________ move to warmer places in winter.
2. Animals that ___________ look as if they are asleep.

Lesson 2

VOCABULARY
hibernate
migrate

How Do Animal Adaptations Help Animals?

Animals that **hibernate** look like they are asleep all winter.

Some animals **migrate** to warmer places for the winter.

Hands-On Activity Observe

1. Watch a fish swimming. Look at the parts of the fish that move. Think about how the fish uses these parts.

   Draw the fish you are watching.

2. What parts of the fish help it move through the water?

   ____

   ____

3. Do you think the parts that help a fish move in water are adaptations? Explain your answer.

   ____

   ____

✓Concept Check

1. A **Cause** is something that makes another thing happen. An **Effect** is the thing that happens. On these pages, find an adaptation that affects an animal's life. Underline the adaptation (cause), and circle its effect.

2. Look on pages 90 and 91. Find two more examples of an adaptation (cause) and its effect. Fill in this graphic organizer.

Adaptation (Cause)	Effect

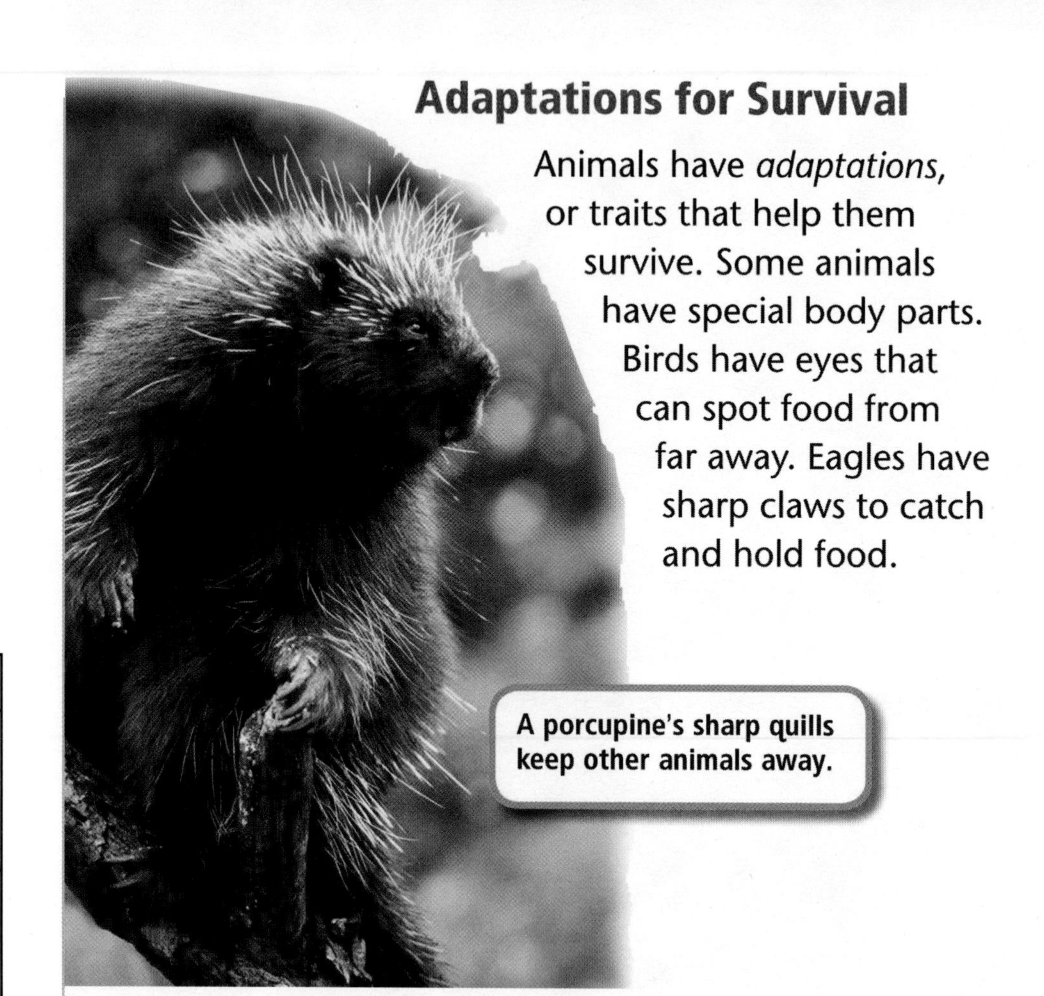

Adaptations for Survival

Animals have *adaptations*, or traits that help them survive. Some animals have special body parts. Birds have eyes that can spot food from far away. Eagles have sharp claws to catch and hold food.

A porcupine's sharp quills keep other animals away.

A crab uses its claws to catch its food.

Bears' sharp claws help them catch fish. They also help the bears climb trees.

Crabs use their claws to catch their food. They also use them to protect themselves.

▼ Bears can use their claws to climb trees.

✓Concept Check

1. Tell one effect of having sharp claws.

____

____

2. The word sharp is used to describe another word. What two words for animal body parts is sharp used with?

____

3. What adaptation for getting food do birds have?

____

4. What do the claws of eagles, bears, and crabs have in common?

____

✓Concept Check

1. A **Cause** is something that makes another thing happen. An **Effect** is the thing that happens. Look for behaviors that cause animals to survive. Underline two behaviors.
2. How does hibernation affect an animal's need for food?

   ____

   ____

3. What is the effect of having colors that match an environment?

   ____

   ____

4. Circle the animal that can hide in different kinds of environments.

Behaviors

Animals have *behaviors*, or ways of acting, that help them. Some behaviors help animals survive winter. There is often little food for animals to find then. So some animals **hibernate**, or go into a kind of deep sleep. That way they need less food. Being able to hibernate is an adaptation.

Other animals **migrate**. Some birds fly south for winter. Knowing how to migrate is an adaptation. It helps the birds survive winter.

▲ When animals hibernate, they use less energy and need less food.

Other Adaptations for Survival

Animals have different adaptations to survive. Some animals hide well. Their colors match those of their environment.

Some animals survive by looking like other animals. Some snakes look like snakes that animals do not like to eat. This keeps them from getting eaten, too.

This animal can change color to match its environment. ▶

Lesson Review

Complete these Cause and Effect sentences.

1. Because some animals ___________ in winter, they use less energy.
2. Because some animals ___________ very well, it is easy for them to spot food.
3. Some animals ___________ to warmer places in winter to find more food.
4. Some animals are hard to see because they match the colors of their ___________.

California Standards in This Lesson

 **3.a** *Students know* plants and animals have structures that serve different functions in growth, survival, and reproduction.

 **3.b** *Students know* examples of diverse life forms in different environments, such as oceans, deserts, tundra, forests, grasslands, and wetlands.

Vocabulary Activity

environment

habitat

climate

Use the vocabulary words above to answer the following questions.

1. Which word means "everything around a living thing"?

____

2. What two things are parts of the answer to question 1?

____

Lesson 3

What Lives in Different Environments?

VOCABULARY
environment
habitat
climate

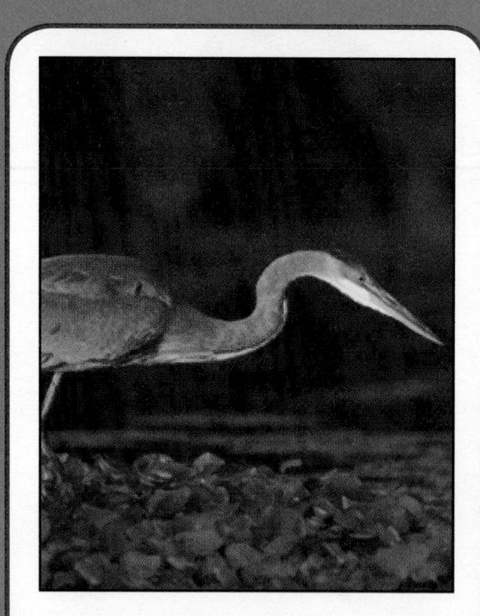

Land, air, and water are all parts of this animal's **environment**.

Some animals live in **habitats** that are under the ground.

Some **climates** have cold winters. Some are warm all year round.

Hands-On Activity
Observe

1. Draw a picture of the environment where you live. Show something about the habitats and climate in your area. Also show some of the plants and animals that live in your area.

2. In what kind of environment do you live?

____

3. What do living things get from their environment?

____

____

✓Concept Check

1. You **Compare** when you look at how things are alike. You **Contrast** when you look at how things are different. What two environments do these pages compare and contrast?

2. Underline the words that tell how living things are able to live in their environments.

Oceans

An ocean is an environment for fish and other plants and animals. An **environment** is everything around a living thing. Water, sun, and seaweed are parts of an ocean environment. Living things have adaptations to live in their environments. Fish have gills that let them breathe in water.

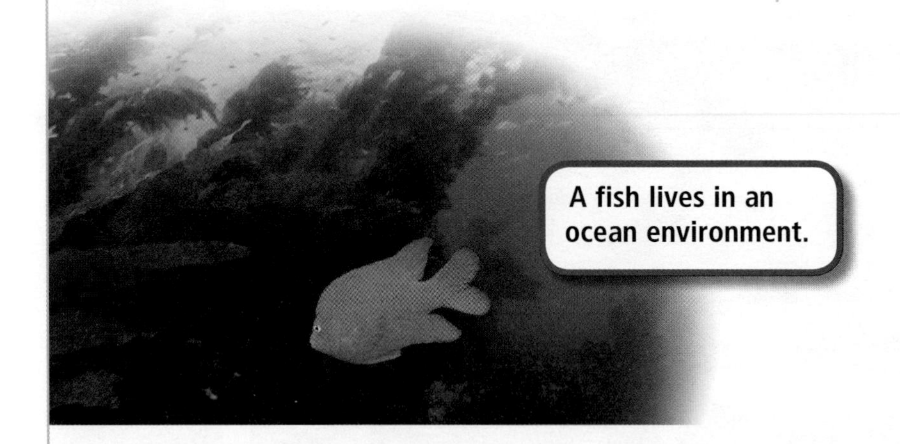

A fish lives in an ocean environment.

Desert Habitats

The desert is a very dry environment. Many deserts are also hot. There are different habitats in any environment. A **habitat** is the part of an environment in which a plant or animal lives.

Some parts of a desert may be flat, with sandy soil. This would be one kind of desert habitat. Another part of the same desert may have rocks and hills. This would be a different habitat.

a desert habitat

✓Concept Check

1. Which is larger—a habitat or an environment?

____

2. Draw an environment that has two different habitats.

3. Give three details about a desert.

____

✓Concept Check

1. You **Compare** when you look at how things are alike. You **Contrast** when you look at how things are different. Look for ways in which the desert is different from the tundra. Name one difference.

____

____

2. How does a desert compare to an ocean?

____

____

3. How would being awake at night help an animal survive in a desert?

____

____

Desert Plants and Animals

Living things in the desert have adaptations to their own habitats. They do not have adaptations to other desert habitats.

Some kinds of cactus live in flat, sandy soil. They survive well in this habitat. But it would be hard for them to live on a rocky hill. They would likely die in this different desert habitat.

This ringtail lives in a rocky desert habitat.

Tundra

Different habitats have different climates. **Climate** is the way the weather is over a long time. Sun, rain, and seasons are part of climate.

The *tundra* has a very cold climate. In some places, the soil is always frozen. Many animals here have have adaptations for living in the cold. The Arctic fox has thick, warm fur. This fur keeps the fox warm in the cold climate. In winter, the fur turns white. It blends in with the snow so the fox can hide.

Animals in the tundra live in cold weather all year.

✓Concept Check

1. Compare and contrast the ways animals survive in cold and hot climates.

Surviving in Cold and Hot Climates

What is the same?	What is different?

2. Underline the words that give the definition of climate.

3. What are two adaptations animals have for living in cold climates?

___

___

✓Concept Check

1. You **Compare** when you look at how things are alike. You **Contrast** when you look at how things are different. Look for ways that forests and grasslands are different. Underline the two sentences that tell the difference.
2. Circle the paragraph that tells about monkeys.
3. Compare the adaptations of a squirrel with those of a monkey.

____

____

____

Forests

Forests are areas with many trees. There are different kinds of forest environments. A tropical rain forest is very hot and wet. Many kinds of monkeys live there.

Monkeys have fingers. This adaptation helps them grab tree branches. They also have long tails that help them keep their balance. This makes it easier for them to move through the trees.

▲ Forests have different habitats for different animals.

Grasslands

Grasslands are flat and covered with grasses. They do not get much rain, but they get more than deserts do.

Grasslands often have fires. The grasses store food in their roots, where it will not burn. After a fire, the grasses quickly grow again.

Most grasslands are flat.

✓Concept Check

1. Compare and contrast grasslands and deserts.

Grasslands and Deserts

The Same	Different

2. What adaptation do grasses have to survive fires?

____

✓Concept Check

1. You **Compare** when you look at how things are alike. You **Contrast** when you look at how things are different. Contrast the body coverings of turtles and frogs.

____

____

2. Compare and contrast wetlands and grasslands.

Wetlands and Grasslands

The Same	Different

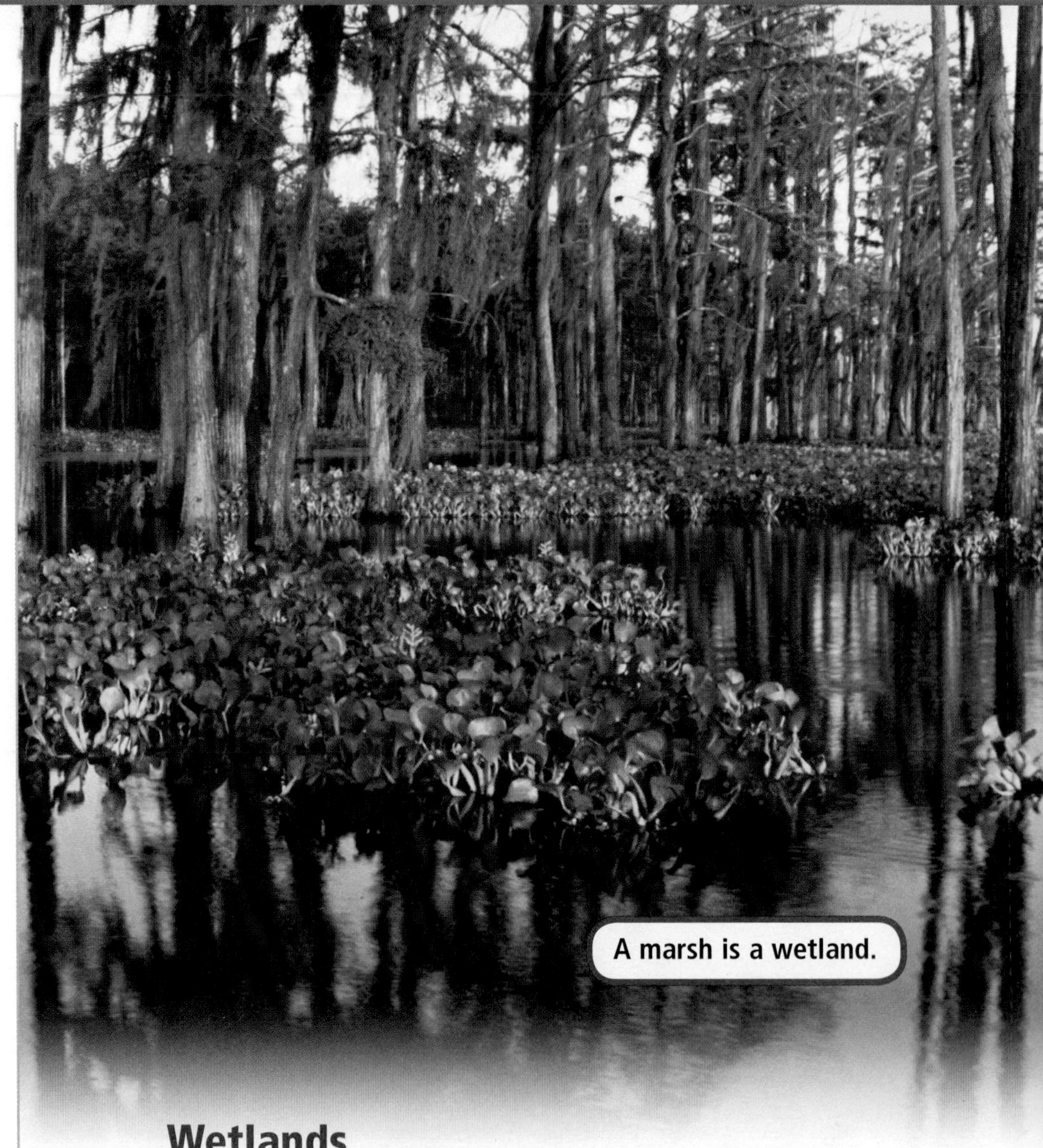
A marsh is a wetland.

Wetlands

Wetlands are areas that are wet most or all of the time. Water animals, such as ducks, live there. Wetland plants need a lot of water. If a wetland dries up, many of the plants die.

Animals in wetlands have adaptations. They help the animals live in a wet enviornment. Turtles have shells for protection. Some ducks can go under the water to get food.

Frogs live in wetlands. ▶

Lesson Review

Complete these Compare and Contrast sentences.

1. An animal's ___________ is smaller than its whole environment.
2. In winter, many ___________ are colder.
3. Some desert habitats are flat and sandy. Others have rocks and ___________.
4. ___________ do not get much rain, but they get more than deserts do.

California Standards in This Lesson

 **3.c** *Students know* living things cause changes in the environment in which they live: some of these changes are detrimental to the organism or other organisms, and some are beneficial.

Vocabulary Activity

1. What is pollution?

____

____

2. Complete the sentence by using these three different forms of the vocabulary word.

   pollutes pollution polluter

   When a ____________ does something harmful to environment, such as making ____________, then he or she _________ it.

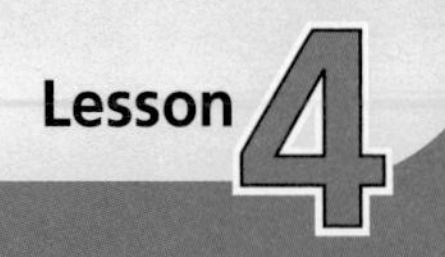

VOCABULARY
pollution

How Do Living Things Change Environments?

Pollution is harmful to the environment. An oil spill is one kind of pollution.

Hands-On Activity
Observe

1. Pour some vegetable oil into a cup of water. Watch the liquids in the cup for a minute or two. Record what happens.

____

2. Now dip your finger into the liquids in the cup. Then remove your finger and look at it. Which liquid sticks to your finger?

____

3. Now put a few drops of water on your oil-covered finger. What happens?

____

4. Birds are sometimes in water. How might oil pollution affect birds?

____

____

____

✓Concept Check

1. When you **Sequence** things, you put them in order. Look for a sequence of events that may harm the environment. What is the first step when people harm the environment? Underline the sentence that tells the first step.

2. What do people do that harms animals that live in trees?

____

3. What is one reason that people change the environment?

____

4. The word harm can be used in different ways in a sentence. Circle two different uses of the word.

People Harm the Environment

People change the environment. Some changes cause harm. People cut down trees. This harms living things that need trees. Animals may lose their homes. Some may die.

People change the environment when they build roads.

People also make **pollution**. Smoke from factories is a kind of air pollution. An oil spill is a kind of water pollution. Animals that get covered with oil may die.

Big machines may take away some animal homes.

✓Concept Check

1. What happens to some animals after trees are cut down? Finish this graphic organizer to show what can happen.

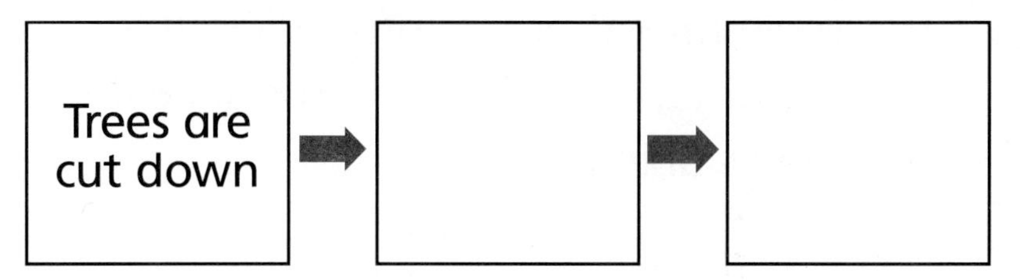

2. Underline the names of two kinds of pollution.

3. When there is an oil spill, what may happen to an animal?

____

4. Where does air pollution come from?

____

✓Concept Check

1. When you **Sequence** things, you put them in order. Look for a sequence of events that may help animals. Some changes to the environment are helpful. Fill in this graphic organizer to show a sequence that helps animals.

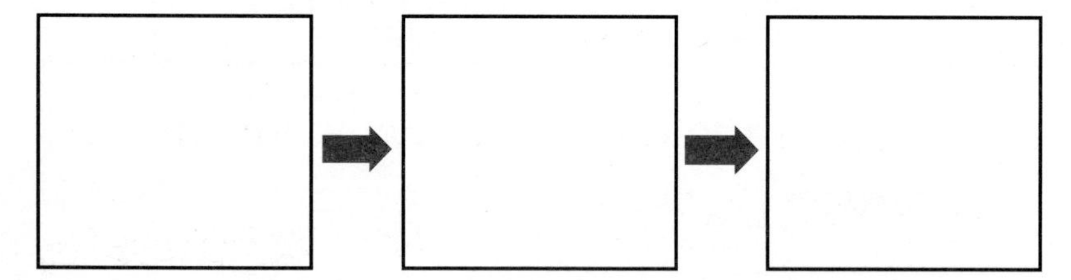

2. After people cut down trees, how can they help the forest grow back?

____

3. What happens when pond plants grow very fast? Finish this graphic organizer to show what can happen.

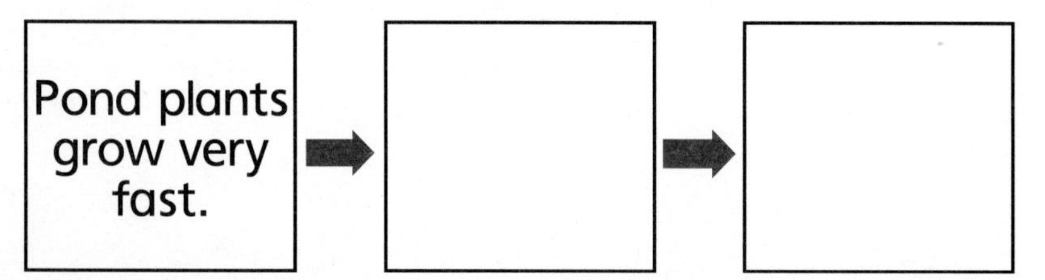

People Help the Environment

Some changes people make can help the environment. After trees are cut down, people can plant new trees. Then forests can grow back faster.

Cars cause air pollution. Special car parts are now made so that cars cause less pollution.

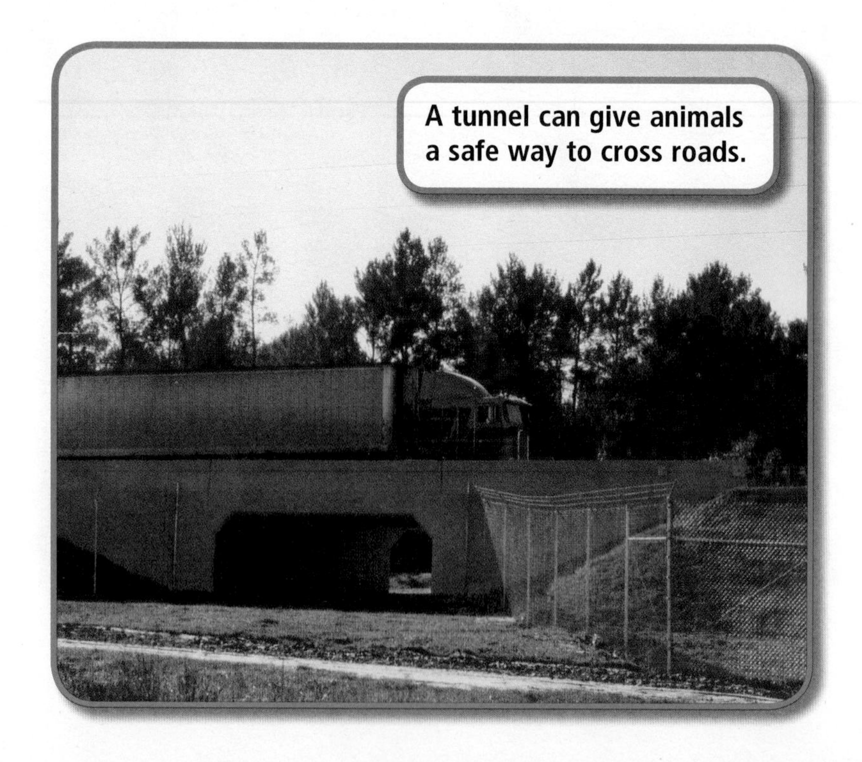

A tunnel can give animals a safe way to cross roads.

Living Things Change Their Environments

Animals change their environments. Some animals dig holes. Beavers build dams to form ponds. The ponds become homes for other animals.

Plants change their environments, too. Some pond plants grow very fast. They block sunlight from plants below. These plants may not survive.

Beavers build dams. ▶

Lesson Review

Complete these Sequence sentences.

1. After trees are cut, many animals must find new ____________.
2. Parts were added to cars, which decreased air ____________.
3. After a beaver builds a dam, a ____________ forms behind it.
4. If pond plants grow very fast, the ____________ cannot reach plants below.

California Standards in This Lesson

 **3.d** *Students know* when the environment changes, some plants and animals survive and reproduce; others die or move to new locations.

Vocabulary Activity

drought

balance

Use the vocabulary words above to complete the following sentences.

1. An environment that has plants and animals in the right amounts is in ____________.

2. In a ____________, there is very little rain for a long time.

Lesson 5

VOCABULARY
drought
balance

How Do Changes to Environments Affect Living Things?

A **drought** is a long time with very little rain.

An environment that is in **balance** has plants and animals in the right amounts.

Hands-On Activity
Experiment, Draw Conclusions

1. Put a thermometer in each of two glass jars with lids. Close the jars. Record the temperature shown by each thermometer.

Kind of Environment	Starting Temperature	Ending Temperature
sunny		
shady		

2. Place one jar in sunlight. Place the other jar in a shady area. Wait 10 minutes. Then record the temperature in each jar.

   What do the jars stand for?

   ____

   ____

Draw Conclusions:

3. How does sunlight change the environment?

   ____

✓Concept Check

1. A **Cause** is something that makes another thing happen. An **Effect** is the thing that happens. Look for examples of a cause and its effect. What cause and effect does page 112 tell about?

____

____

2. Find two more examples of a cause and its effect on page 113. Fill in this graphic organizer.

Cause and Effect

The Cause	The Effect

Climate Affects Living Things

Different environments have different climates. Animals have adaptations that help them survive in their climates. During a **drought**, there is little rain for a long time. Some plants are used to getting a lot of water. During a drought, these plants may die.

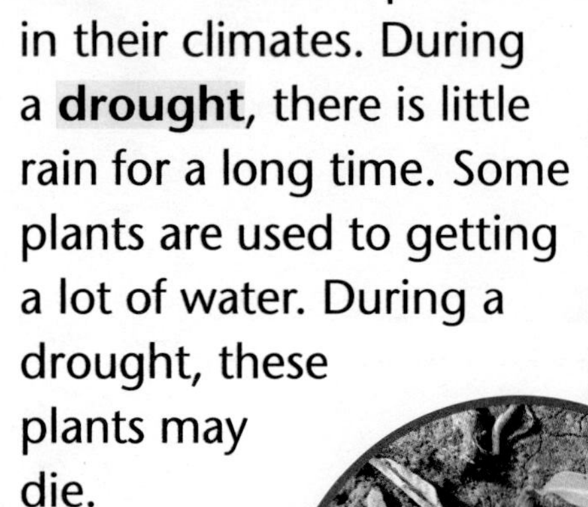

Plants may die in a drought. ▶

▲ Water helps people stay cool in hot climates.

The desert is hot and dry. Few animals can live there. Those that do have special adaptations. Desert plants grow very fast when it rains. They make seeds quickly before the land dries up again.

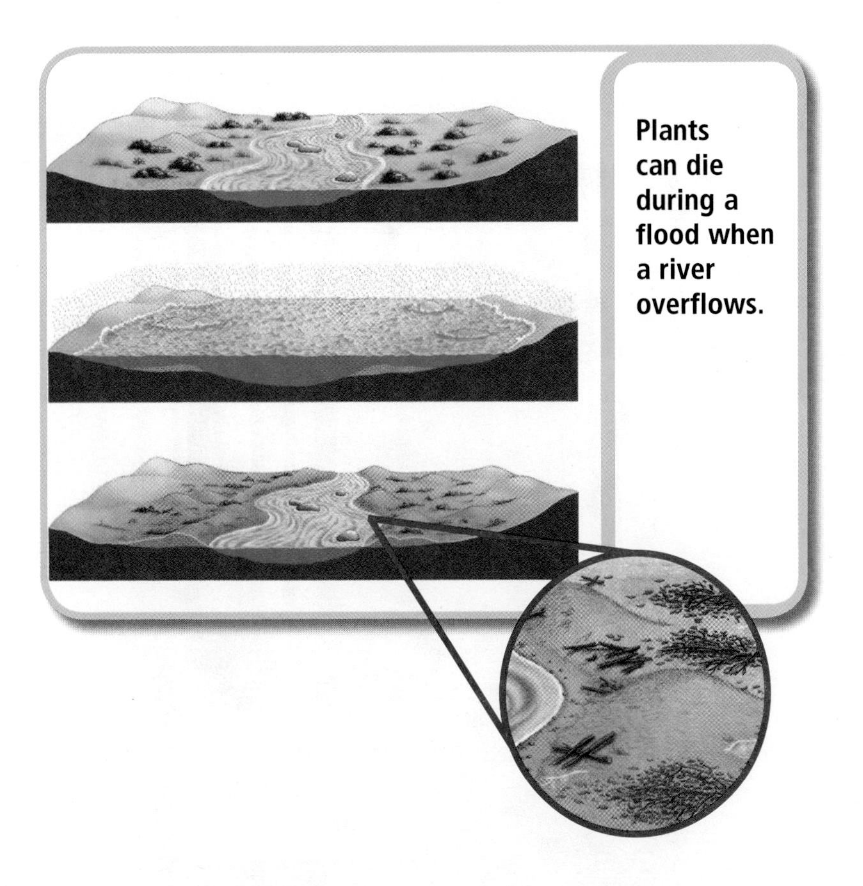

Plants can die during a flood when a river overflows.

✓Concept Check

1. What effect can a drought have on plants?

____

2. Circle two words that give details about the desert.

3. Look at the pictures on page 113. What caused the change shown in the third picture?

____

4. What causes desert plants to grow very fast?

____

✓Concept Check

1. A **Cause** is something that makes another thing happen. An **Effect** is the thing that happens. What caused the changes shown on page 114?

____

2. How can fires cause some plants to grow better?

____

____

3. What is the effect of having too few plants in an environment?

____

____

Fire Affects Living Things

Forest fires change the environment. Many trees and other plants burn. Some animals lose their homes. If they cannot find new homes, they may die.

Fires can be good for forests, too. Burned trees often fall. This lets the sun reach small plants on the ground. More light helps them grow better.

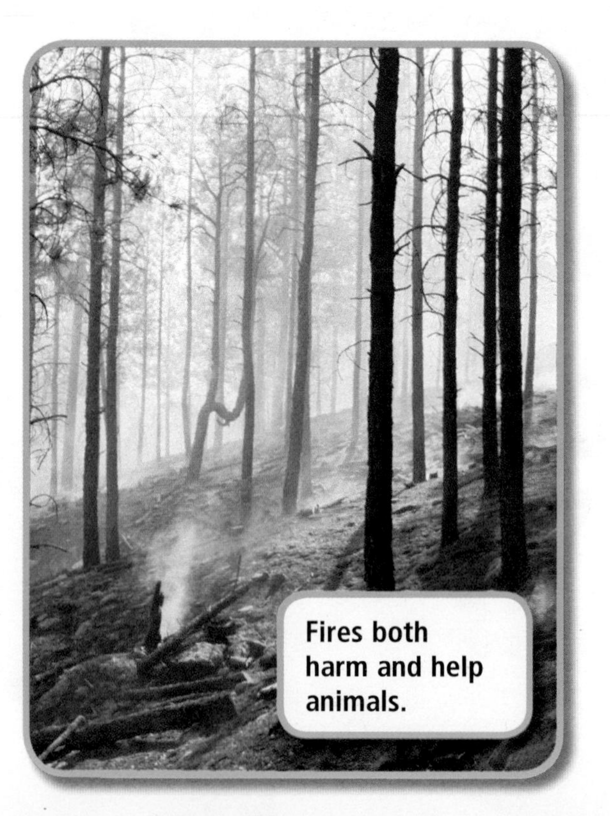

Fires both harm and help animals.

Balance Between Living Things

Balance means having not too many and not too few of each thing. Environments need the right numbers of the right plants. Without them, animals will not have the foods they need. They will have to move or die.

This pond environment is in balance. It has plants and animals in the right amounts.

A pond environment in balance

Lesson Review

Complete these Cause and Effect statements.

1. A ___________ causes plants to dry up.
2. When plants dry up, they often ___________.
3. A drought can cause forest ___________.
4. Fires help keep the ___________ of not too many of one kind of plant or animal.

California Standards in This Lesson

 **3.e** *Students know* that some kinds of organisms that once lived on Earth have completely disappeared and that some of those resembled others that are alive today.

Vocabulary Activity

fossil **extinct**

Use the vocabulary words above to answer the questions below.

A 

B 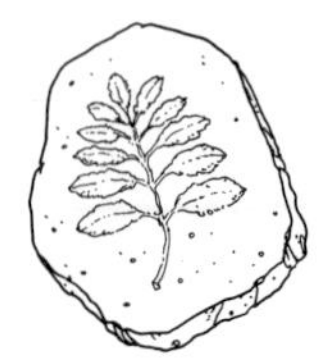 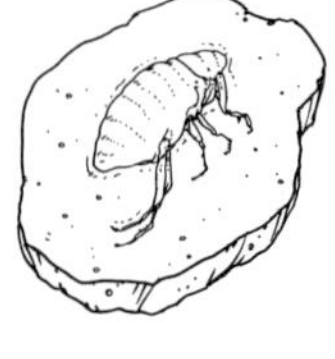

1. Which vocabulary word tells about picture B?

___

2. Which vocabulary word tells about picture A?

___

Lesson 6

VOCABULARY
fossil
extinct

How Have Living Things Changed over Time?

Fossils help us learn about plants and animals that lived long ago.

Dinosaurs died out a long time ago. They became **extinct**.

Hands-On Activity
Observe, Interpret

1. Find a picture of a fossil. Draw the fossil in the box below.

2. What plant or animal alive today does the fossil most look like?

____

3. How is the fossil the same as and different from the plant or animal?

____

____

____

____

✓Concept Check

1. You **Compare** when you look at how things are alike. You **Contrast** when you look at how things are different. Compare and contrast extinct and living plants and animals. How are they alike, and how are they different?

Alike	Different

2. Underline two details about fossils on page 118.

Animals Then and Now

Fossils are remains of plants or animals that died long ago. They are very hard. Some fossils are animal parts, such as bones or teeth. Others are just prints left in mud.

This fish was thought to be extinct.

Fossils show how animals have changed. No animals today look like dinosaurs. Camels today look much the same as camels long ago but are larger.

The fish was found to still be living today.

✓Concept Check

1. Compare camels of today to extinct camels.

___

2. Compare the two fish in the pictures on this page.

___

___

___

3. Contrast the two fish in the pictures on this page.

___

___

4. What do the pictures of the fish on this page tell us?

___

___

✓Concept Check

1. You **Compare** when you look at how things are alike. You **Contrast** when you look at how things are different. Contrast animal fossils and plant fossils.

___

___

2. Compare plant fossils to animal fossils.

___

___

___

3. Underline the two sentences on page 121 that contrast what different animals do when climates change.

Plants Then and Now

There are more animal fossils than plant fossils. Plants are soft and do not last well. Some plants left leaf prints in mud. The prints became fossils. Some fossils were once tree trunks that became like stone.

Fossils show that some kinds of plants no longer grow on earth. Other fossils look like plants that still live today.

A fern fossil

Extinct Plants and Animals

▲ The woolly mammoth lived during the Ice Age.

Extinct plants and animals no longer live on Earth. When the last one died, there were no more. All dinosaurs are gone from the earth.

Living things become extinct for different reasons. Sometimes climates change. Some animals are able to adapt and survive. Some are not. During the Ice Age, many kinds of animals died out. They could not adapt to living in the cold.

Lesson Review

Complete these Compare and Contrast sentences.

1. ____________ are very hard, like rocks.
2. Some fossils are animal parts. Others are ____________ left in the mud.
3. There are more ____________ fossils than plant fossils.
4. Some plants leave leaf prints in ____________.

Circle the letter in front of the best choice.

1. Which adaptation keeps rose plants from being eaten?

   A flowers
   B thorns
   C leaves
   D stems

2. Which of these plants has an adaptation for growing up a tree?

   A flytrap
   B rosebush
   C sunflower
   D vine

3. Which of these adaptations is a kind of behavior?

   A claws
   B quills
   C migrating
   D changing color

4. A plant needs to grow in shade to survive. In which environment would the plant most likely live?

   A a forest
   B a tundra
   C a grassland
   D a desert

5. How do beavers help other animals by changing the environment?

   A by cutting down trees
   B by hibernating in winter
   C by migrating to different streams
   D by building dams that make ponds

6. Which of these is true of all pollution?

   A It is helpful to some animals.
   B It is harmful to living things.
   C It is made when plants burn.
   D It is found on dry land.

7. What is a drought?

A a long period with little rain

B a habitat that is very dry

C a change in the climate

D an environment that is mostly water

8. What must any environment have for its living things to survive?

A only plants

B only animals

C a balance of plants and animals

D a fire that kills some of its trees

9. Which is true of all fossils?

A They are remains of plants.

B They are remains of animals.

C They are remains of things that died long ago.

D The things that made them still live on Earth.

10. List two kinds of pollution. Give an example of each kind.

a. ______________________

b. ______________________

11. List three adaptations that help animals survive.

a. ______________________

b. ______________________

c. ______________________

12. Look back at the question you wrote on page 80. Do you have an answer for your question? Tell what you learned that helps you understand the adaptations living things have to survive in their environments.

____

____

____

____

Patterns in the Sky

In this unit, you will learn about the planets that orbit the sun, the patterns followed by Earth and the sun and stars, and why the shape of the moon seems to change. What do you know about these topics? What questions do you have?

Thinking Ahead

What planets orbit the sun? Draw an arrow to show where Earth is in this picture of the solar system.

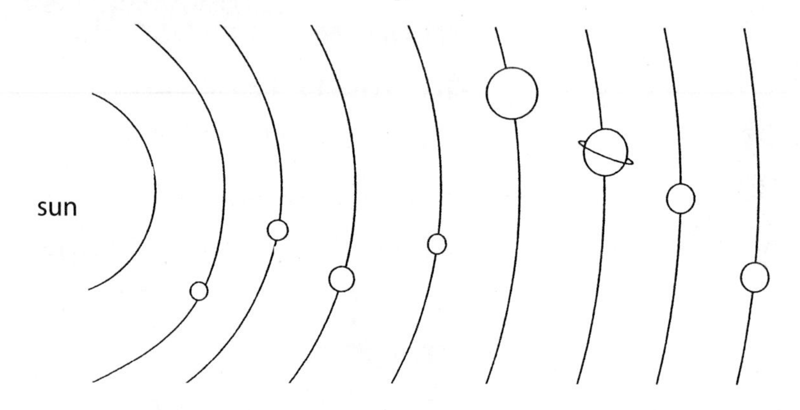

What patterns do the stars follow?

____

____

Why does the shape of the moon seem to change?

____

____

____

Write a question you have about the patterns followed by Earth and the sun.

____

____

Recording What You Learn

On this page, record what you learn as you read the unit.

Lessons 1 and 2

What planets orbit the sun?

___

___

Lesson 3

What is a constellation?

___

___

Lesson 4

Why does the shape of the moon seem to change? Name each phase of the moon.

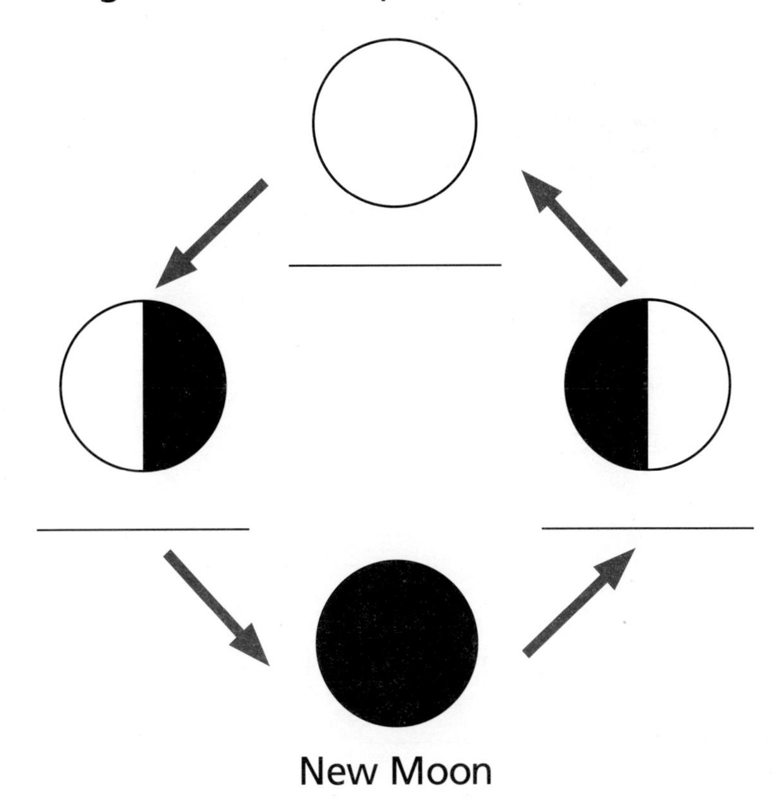

California Standards in This Lesson

4.c *Students know* telescopes magnify the appearance of some distant objects in the sky, including the moon and the planets. The number of stars that can be seen through telescopes is dramatically greater than the number that can be seen by the unaided eye.

4.d *Students know* that Earth is one of several planets that orbit the sun and that the moon orbits Earth.

Vocabulary Activity

moon **planet**

orbit **telescope**

1. Use three of the vocabulary words above in a sentence.

____

____

2. A tool that is used to study the solar system is a ____________.

Lesson 1

VOCABULARY

planet
orbit
telescope
moon
solar system

What Planets Orbit the Sun?

A **planet** is a large object in space that travels around a star. This is the planet Saturn.

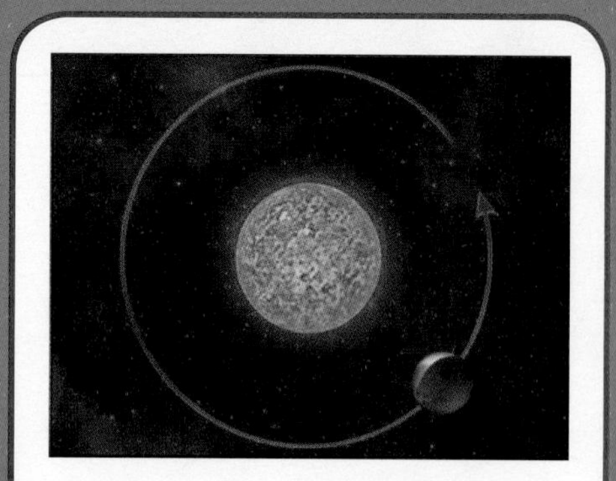

To travel around an object is to **orbit**. This planet is orbiting the sun.

A **telescope** is a tool that scientists use to see objects that are far away. Objects seem closer and larger when you look at them through a telescope.

The **solar system** is the sun and everything that orbits it. The solar system includes nine planets and the planets' moons.

A **moon** is a large object that orbits a planet. The Earth has one moon that orbits it.

Hands-On Activity
Make a Model

1. Tie the loose end of the thread on a spool to a key ring. The key ring stands for the sun. Use buttons of different sizes to show the planets of the solar system. Choose a button for each planet to show how it is larger or smaller than the other planets.

Which planet will have the largest button?

____

Which planet will have the smallest button?

____

2. Tie the buttons to the thread to model the solar system. Be sure the planets are in the same order as shown in the picture.

How is your model the same as the solar system in the picture?

____

____

✓Concept Check

1. You **Compare** when you look at how things are alike. You **Contrast** when you look at how things are different. Underline two ways planets are alike.
2. What is the sun?

____

3. Circle the two different forms of the word orbit on these pages.
4. Are all the planets moving all the time?

____

Planets and Moons

We live on Earth. Earth is a planet. A **planet** is a large body of rock or gas in space. Planets are moving. Earth is moving right now, but you cannot feel it. Planets **orbit**, or travel around, a star. Earth is one of nine planets that orbit the sun.

The moon is smaller than Earth.

A **moon** is a large body that orbits a planet. Earth has one moon that orbits it. The moon takes about one month to orbit Earth once. You can see the moon in the sky at night.

✓Concept Check

1. How is a moon different from a planet?

____

____

2. Draw a planet and its moon.

3. Contrast the time it takes the moon to orbit Earth with the time it takes Earth to orbit the sun.

____

____

✓Concept Check

1. You **Compare** when you look at how things are alike. You **Contrast** when you look at how things are different. Compare and contrast the planets of the solar system. Fill in this graphic organizer.

Planets of the Solar System

How are they alike?	How are they different?

2. How do the planets move in the solar system?

____

3. Where in the solar system is the sun found?

____

The Solar System

The **solar system** is the sun and the things that orbit it. The sun is the center of the solar system. The word *solar* means "sun."

There are nine planets in the solar system. The moons that orbit the planets are part of the solar system, too.

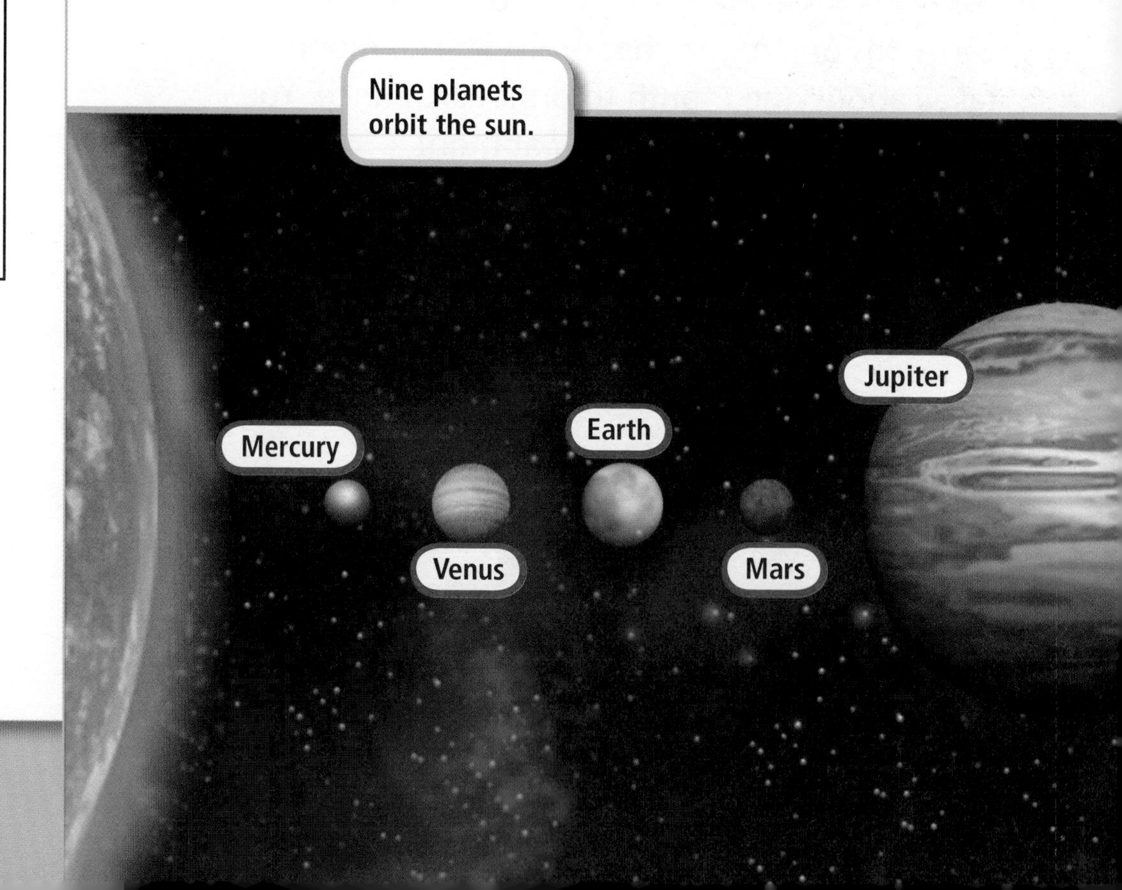

Looking at the Sky

A **telescope** is a tool that helps you see things that are far away. Telescopes make things look closer and larger. A telescope helps you see details. You need a telescope to see the planets that are far away.

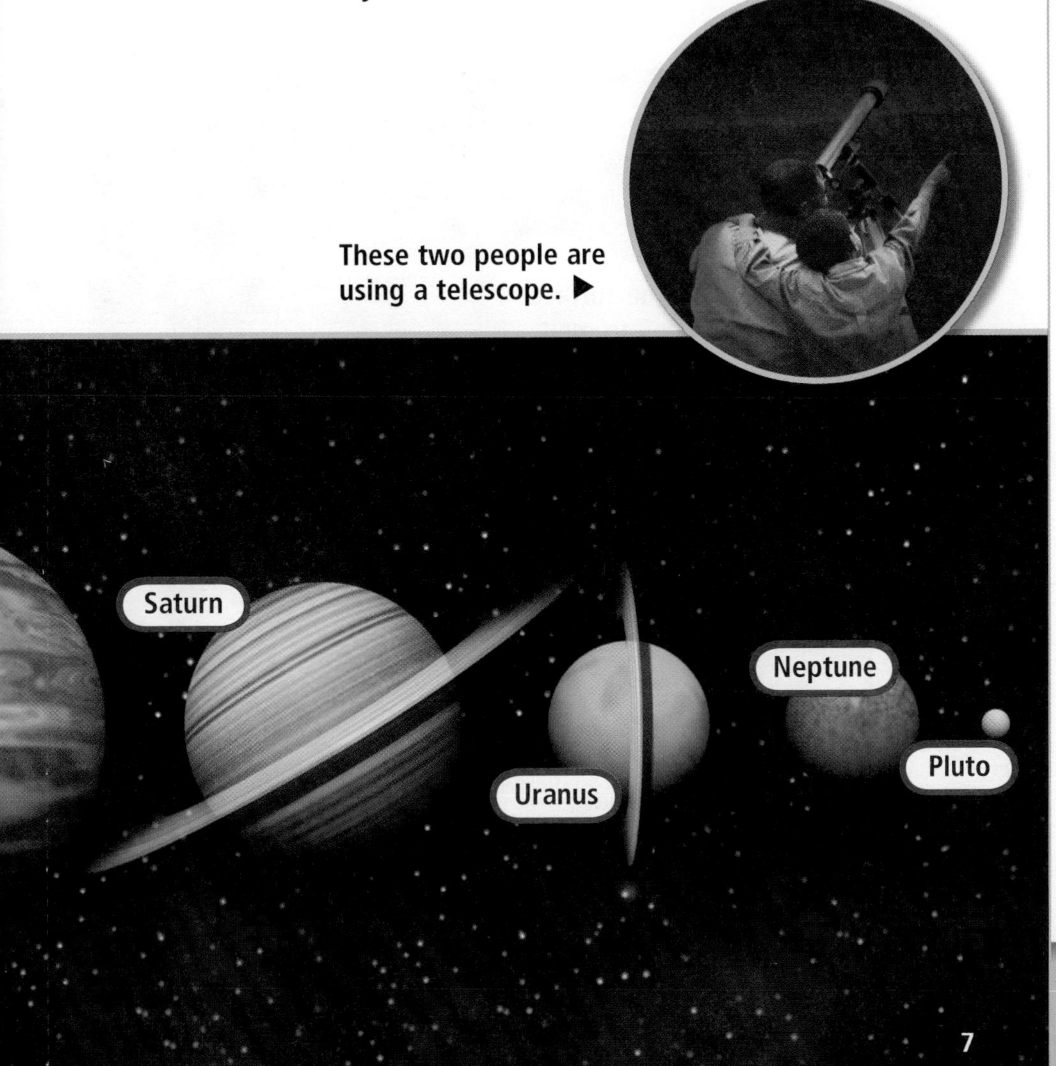

These two people are using a telescope. ▶

✓Concept Check

1. How is looking through a telescope different from using your eyes alone?

____

____

____

____

2. Underline the sentence that tells what a telescope is.

3. Do you think you could see the rings around Saturn and Uranus with only your eyes? Explain your answer.

____

____

____

✓Concept Check

1. You **Compare** when you look at how things are alike. You **Contrast** when you look at how things are different. How are the inner planets alike?

____

____

____

2. What is one way to compare Mercury and Venus?

____

3. Contrast the sizes of Mercury and Venus.

____

The Inner Planets

There are nine planets in the solar system. The four planets that are closest to the sun are called the inner planets. The inner planets are Mercury, Venus, Earth, and Mars.

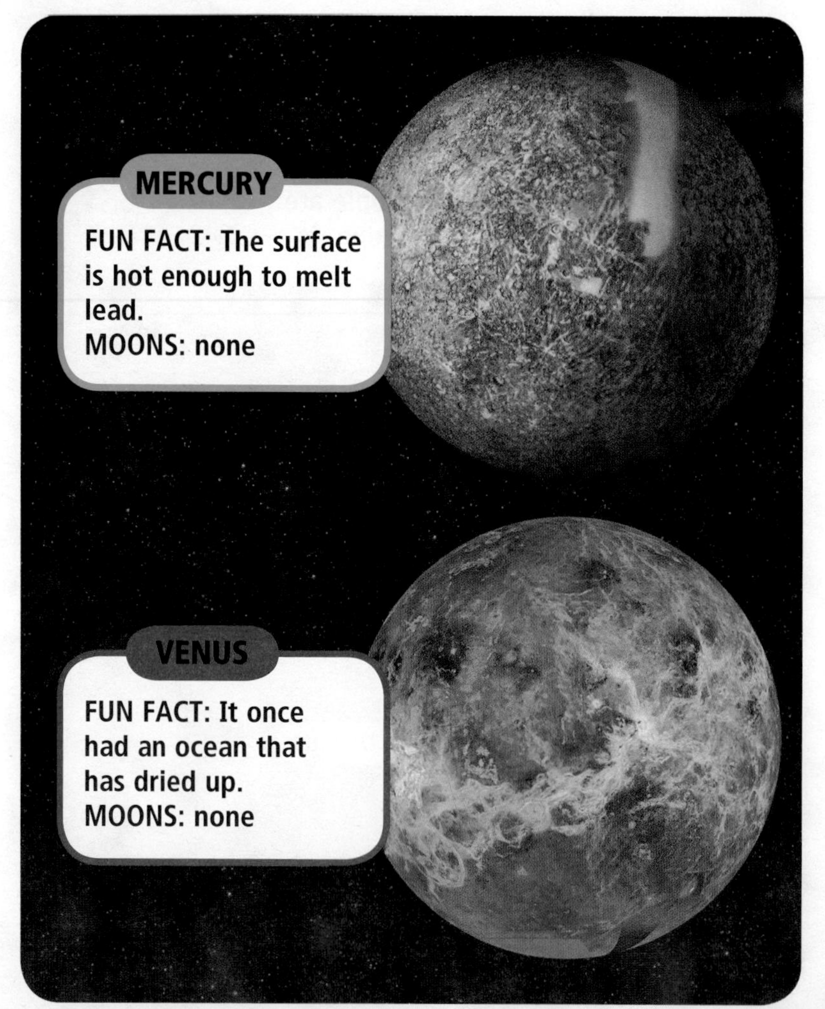

All the inner planets have rocky surfaces. They are also warmer than some of the other planets because they are closer to the sun.

✓Concept Check

1. How is Earth different from the other inner planets?

____

____

2. Look back at page 130. Contrast the sizes of Earth and Mars by drawing two circles in the box below.

Sizes of Earth and Mars

Earth	Mars

3. Which inner planet has the most moons?

____

4. Which inner planet has the largest volcano?

____

✓Concept Check

1. You **Compare** when you look at how things are alike. You **Contrast** when you look at how things are different. Compare and contrast the outer planets. Which outer planet is most different from the others? Explain your answer.

____

____

2. What is one way the outer planets are different from the inner planets?

____

____

3. How are Saturn and Uranus different from the other outer planets?

____

4. How is Uranus different from the other outer planets?

____

The Outer Planets

The five planets that are farthest from the sun are called the outer planets. They are Jupiter, Saturn, Uranus, Neptune, and Pluto. Most of them are larger than the inner planets. The outer planets are made of frozen gases.

NOTE: In 2006, scientists from the International Astronomical Union classified Pluto as a "dwarf planet." However, not all scientists agree with the decision. Use the Internet or current newspaper and magazine articles to find the latest information about the planets.

Lesson Review

Complete these sentences that Compare planets.

1. Nine planets ____________ the sun.
2. All ____________ are large bodies of rock or gas that travel around the sun.

Complete these sentences that Contrast planets.

3. The ____________ planets are closer to the sun than the ____________ planets.
4. Some planets have one or more ____________ orbiting them.

California Standards in This Lesson

 **4.e** *Students know* the position of the sun in the sky changes during the course of the day and from season to season.

Vocabulary Activity

rotation

axis

1. Use the vocabulary words above correctly in a sentence.

____

____

2. What is the Northern Hemisphere?

____

____

____

Lesson 2

VOCABULARY
rotation
axis
Northern Hemisphere

What Patterns Do Earth and the Sun Follow?

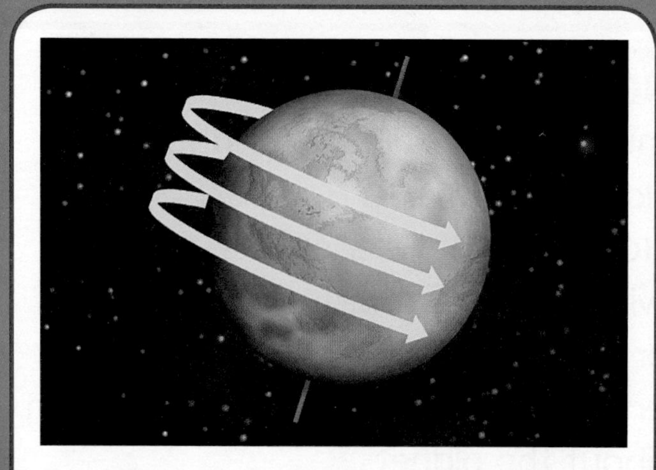

The Earth is always spinning. The spinning is called **rotation**. One rotation of Earth takes 24 hours.

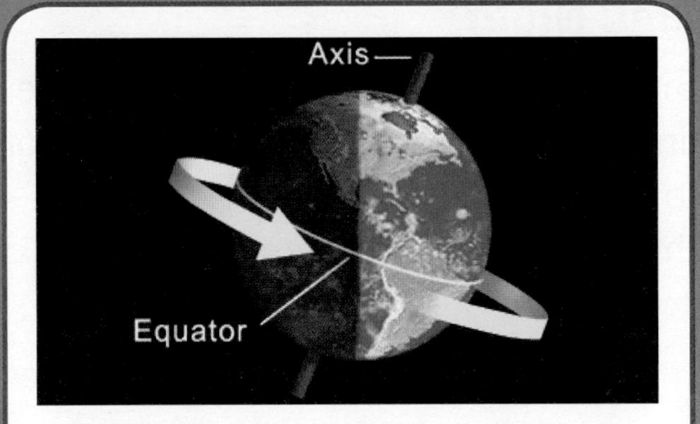

Earth has an imaginary line that runs through it. The line runs from the North Pole to the South Pole. We call this the Earth's **axis**. Earth spins around its axis.

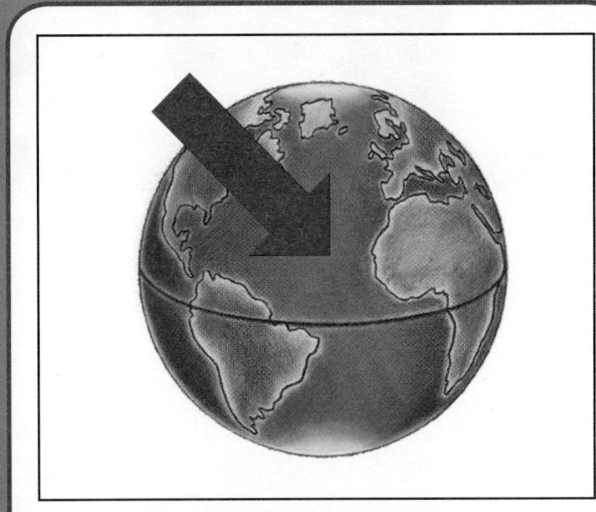

The **Northern Hemisphere** is the northern half of Earth.

Hands-On Activity
Observe

1. Find one large ball to stand for the sun and a smaller ball to stand for Earth. Use them to explain to a family member some things you learned about space.
2. How could you use the two balls to show how Earth orbits the sun?

____

____

3. How could you use the smaller ball to show how Earth rotates?

____

____

✓Concept Check

1. A **Cause** is something that makes another thing happen. An **Effect** is the thing that happens. Look for the effects of the movements of Earth. What is one effect of the rotation of Earth?

____

2. What is another word that means the same thing as rotating? Circle this word.

3. How many hours does it take Earth to make one rotation?

____

4. How many rotations will Earth make in one year?

____

Day and Night

Earth moves in two ways. **Rotation** is the spinning of Earth. It takes Earth 24 hours, or one day, to make one rotation.

Earth also orbits the sun. It takes Earth 365 days to orbit the sun.

The sun is lighting only the side of Earth that is facing it. This is why we have day and night. The part of Earth that is lit by the sun is always changing because of Earth's rotation. When it is daytime for one half of Earth, it is night for the other half.

◀ **The sun is shining on California.**

At the same time, it is night on the other half of the Earth. ▶

✓Concept Check

1. What causes day and night?

____

____

2. Draw a picture in the box below. Show the sun and Earth and where it is day and night.

✓Concept Check

1. A **Cause** is something that makes another thing happen. An **Effect** is the thing that happens. Look for the effects of the movements of Earth. What is one effect of the sun's changing position in the sky?

____

____

2. What causes shadows to get longer and shorter?

____

____

3. You see the sun rising. In what direction are you looking?

____

Shadows Change

Every day, the sun comes up in the east. Later, the sun sets in the west. Notice how the sun's position changes in the pictures on the next page. Shadows change because the sun's position changes.

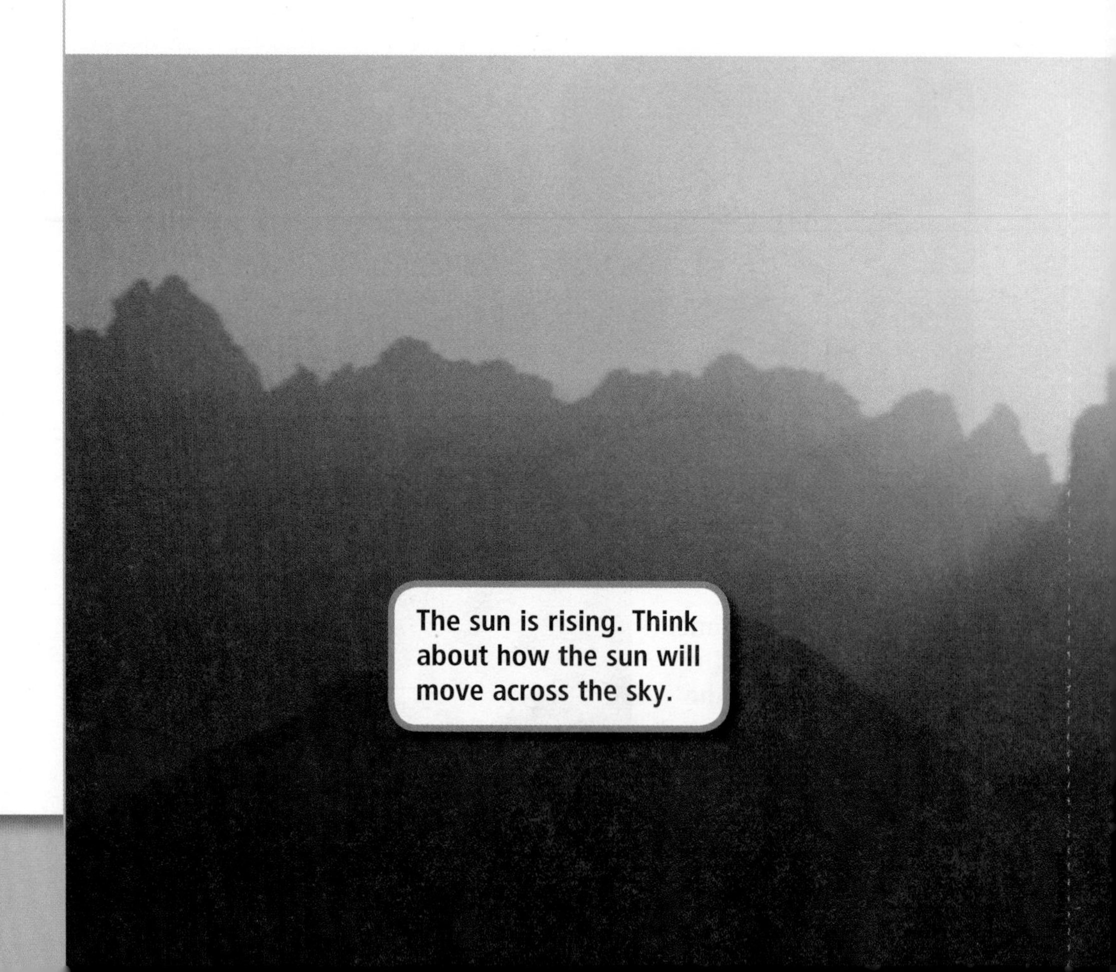
The sun is rising. Think about how the sun will move across the sky.

The shadows show us how this part of Earth has rotated away from the sun. ▶

✓Concept Check

1. The pictures show a beach at different times of the day. On which side of the umbrella does its shadow form?

____

2. The sun moves from east to west across the sky. How do the shadows move?

____

3. Where would the sun be in the sky when the shadows are shortest?

____

____

✓Concept Check

1. A **Cause** is something that makes another thing happen. An **Effect** is the thing that happens. What is the cause of summer and winter?

____

____

2. What part of Earth has winter?

____

3. The sun gets higher in the sky in summer. What is the effect of the movement?

____

4. Draw an arrow to Earth's axis in one of the pictures on page 143.

The Sun's Positions

Earth has an imaginary line that runs through it. The line is called its **axis**. Earth's axis is not straight up and down. It is tilted toward or away from the sun. The part of Earth that is tilted toward the sun has summer. The part that is tilted away from the sun has winter.

California is in the **Northern Hemisphere**, the half of Earth that is closer to the North Pole. In summer, the sun is higher in the sky, so the days are longer. In winter, the sun is lower in the sky and the days are shorter.

In summer, the sun is higher in the sky. ▼

▲ In winter, the sun is lower in the sky.

The Northern Hemisphere is tilted away from the sun. It has winter. ▶

◀ The Northern Hemisphere is tilted toward the sun. It has summer.

Lesson Review

Complete these Cause and Effect statements.

1. The sun seems to move across the sky because of the __________ of Earth.
2. The tilt of Earth's __________ causes winter and summer.
3. California is in the __________ __________. This is the northern half of Earth.
4. Earth __________ the sun. This takes 365 days.

California Standards in This Lesson

 **4.a** *Students know* the patterns of stars stay the same, although they appear to move across the sky nightly, and different stars can be seen in different seasons.

 **4.c** *Students know* telescopes magnify the appearance of some distant objects in the sky, including the moon and the planets. The number of stars that can be seen through telescopes is dramatically greater than the number that can be seen by the unaided eye.

Vocabulary Activity

constellation

magnify

Use the vocabulary words above to complete the following sentences.

1. A pattern that is seen in stars is called a ____________.

2. Telescopes ____________ stars to make them look larger.

What Patterns Do Stars Follow?

VOCABULARY
star
constellation
magnify

When you look up in the night sky, you can see **stars**. They are very far away.

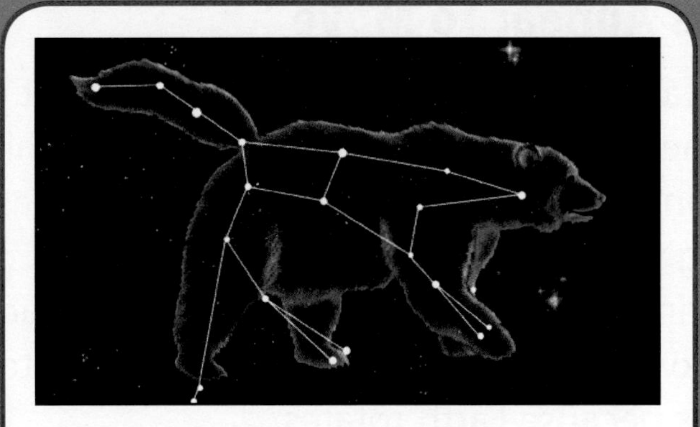

If you connect this group of stars in the sky, you can make a picture of a bear. The stars that you connect form a **constellation**.

When you **magnify** something, you make it seem larger. The person is using a hand lens to magnify the apple.

Hands-On Activity
Observe

1. On a clear night, look out a window and find a constellation. Draw the constellation.

2. Wait three hours. Find the same constellation you saw before. Is it in the same position in the sky?

____

3. Write about any changes you saw.

____

____

✓Concept Check

1. The **Main Idea** on these two pages is Stars seem to move. **Details** tell more about the main idea. Why do stars seem to move? Underline the sentence on page 146 that gives this detail.

2. Look for details about stars. Fill in this graphic organizer to show what you found.

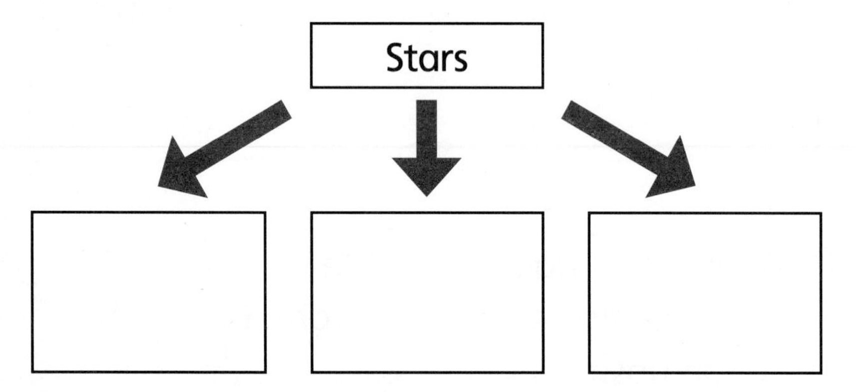

3. Look at the pictures. Notice the three stars in a row. How did these stars seem to move?

____

____

Stars Appear to Move

A **star** is a ball of hot, glowing gases that gives off energy. You can see stars at night as white dots in the sky. Those dots are really hot gases that give off light.

During the day, the sun seems to move across the sky. At night, the stars seem to move, too. This is because Earth rotates.

▲ **The student watches the stars at different times at night. They seem to move across the sky.**

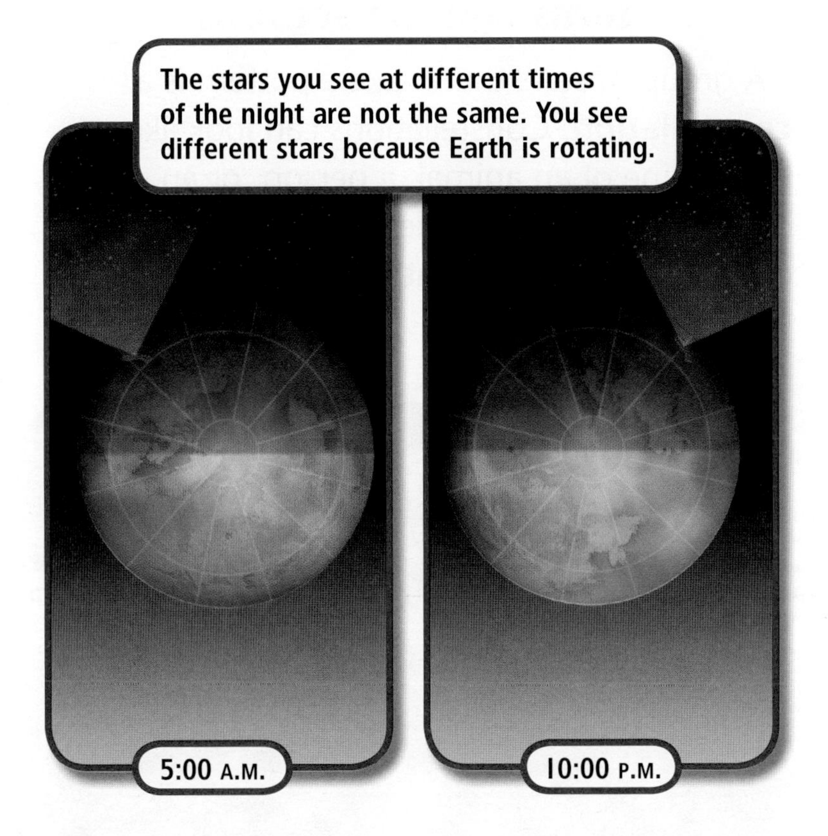

The patterns of stars we see in the sky do not change. The Big Dipper is a pattern of stars in the sky. It always looks the same. But it is not always in the same place in the sky. Since Earth is rotating, the stars look as if they are moving.

✓Concept Check

1. What is a star?

____

____

2. What does not change when we look at the sky?

____

____

3. Do the star patterns stay in the same place?

____

____

✓Concept Check

1. The **Main Idea** on these two pages is Constellations are star patterns. **Details** tell more about the main idea. Look for details in the pictures at the bottom this page. What is one detail that stays the same from one picture to the next?

____

____

2. What are the names of the four seasons of the year?

____

3. Underline a sentence that tells what constellations can look like.

4. What is the name of one constellation?

____

Star Patterns and the Seasons

A group of stars that make a pattern is a **constellation**. Constellations can look like the outline of an animal, a person, or an object. Long ago, people gave names to these constellations. The Great Bear is a constellation.

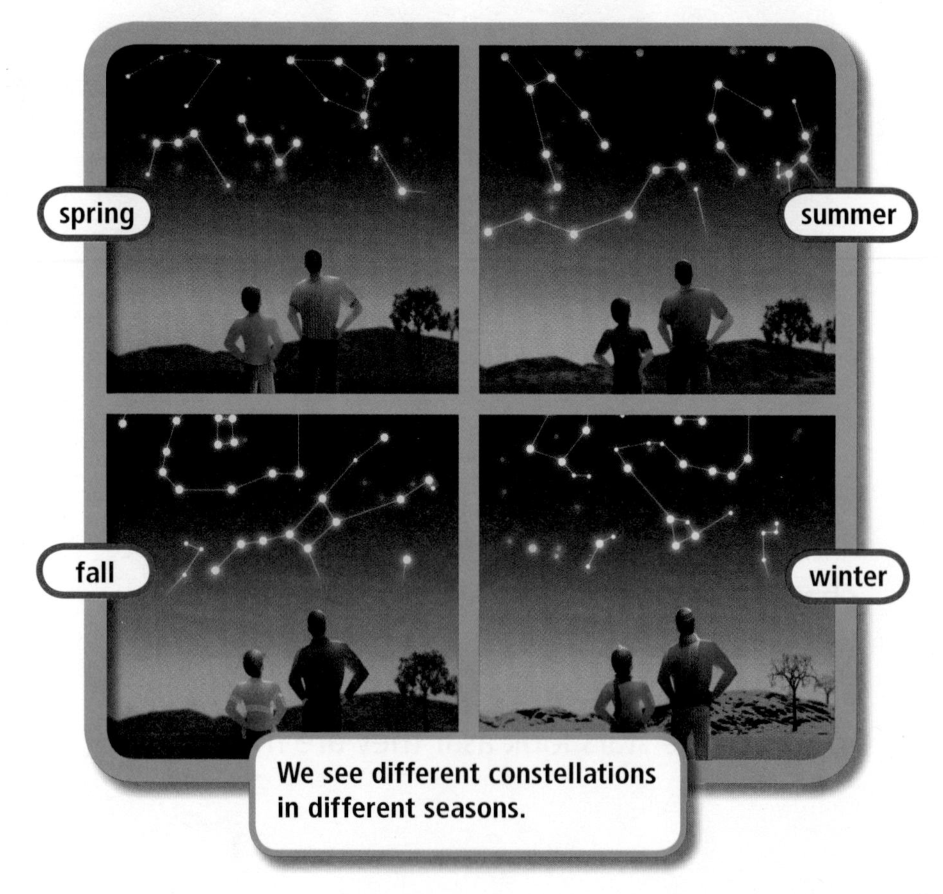

We see different constellations in different seasons.

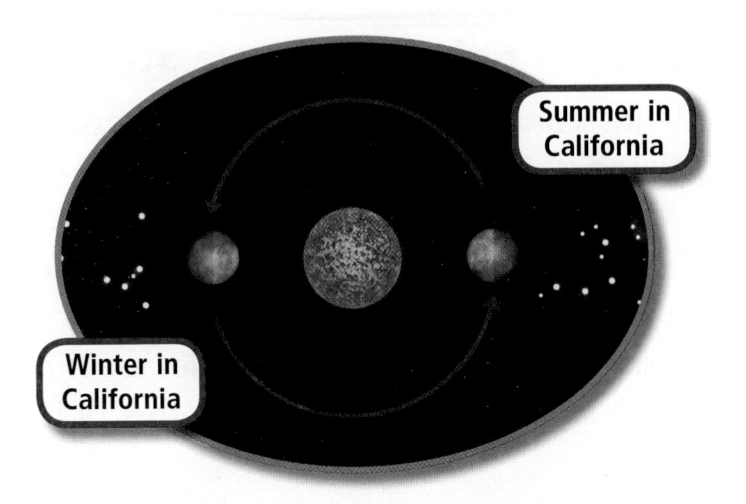

The stars you can see depends on where Earth is in its trip around the sun. As Earth orbits the sun, the seasons change. You can see different stars each season.

✓Concept Check

1. Why do we see different star patterns in different seasons?

____

____

____

2. Look at the picture at the top of the page. How is Earth's position in its orbit different in summer and in winter?

____

____

3. What is the name of the constellation in the picture at the bottom of the page?

____

4. What happens as the seasons change on Earth?

____

✓Concept Check

1. The **Main Idea** on these two pages is Telescopes help us observe stars. **Details** tell more about the main idea. How do scientists use what they see with telescopes?

____

____

2. What can we see with a telescope that we can't see with our eyes alone? List three things in the chart below.

Things Telescopes Help Us See

3. Look at the picture on page 151. Circle two words that give details about the telescopes shown in the picture.

Telescopes Help Us Observe Stars

Scientists and other people use telescopes to see the stars. Telescopes help us see things that are very far away. You can see more stars through a telescope than if you were just looking up at the sky. A telescope will **magnify** the stars, or make them look larger. Telescopes help scientists study stars, planets, and other objects in the sky.

On Earth, there are many very large telescopes. They are used to see details of objects in space. Scientists use them to learn more about the solar system.

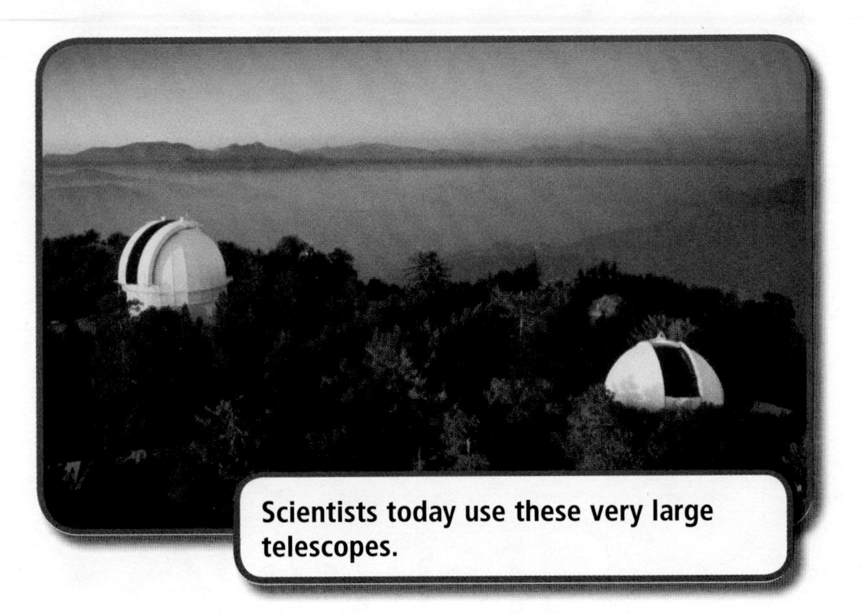

Scientists today use these very large telescopes.

Lesson Review

Complete this Main Idea sentence.

1. The ___________ appear to move because Earth is rotating.

Complete these Detail sentences.

2. We see different star patterns in ___________ and ___________.

3. A ___________ is a group of stars that look like a certain pattern.

4. A ___________ helps us observe stars.

California Standards in This Lesson

 **4.b** *Students know* the way in which the moon's appearance changes during the four-week lunar cycle.

Vocabulary Activity

A student labeled these pictures of the moon.

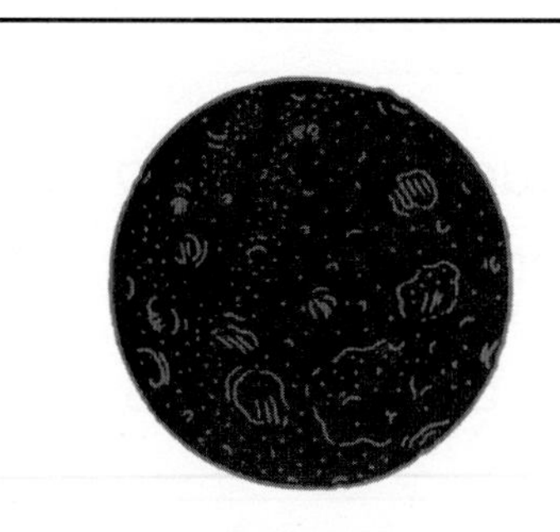	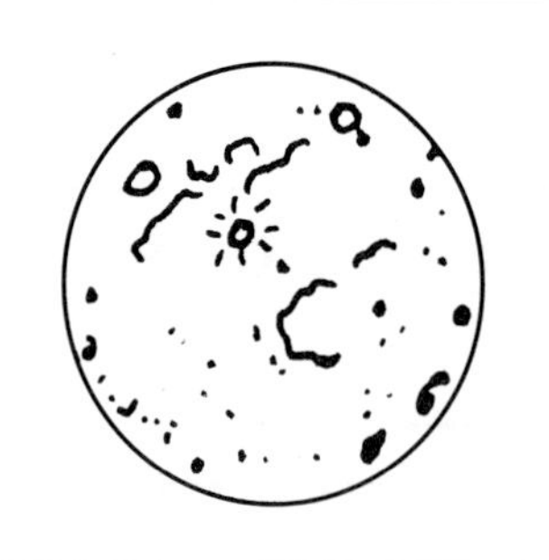

New Moon | Full Moon

1. Did the student correctly label these phases of the moon?

___

2. If the labels are not correct, how should they be changed?

___

Lesson 4

Why Does the Shape of the Moon Seem to Change?

VOCABULARY
moon phases
full moon
new moon

The moon seems to have different shapes as it orbits Earth. The different shapes are called **moon phases**.

You see a **full moon** when the moon looks like a circle.

When the whole moon looks dark, it is a **new moon**. The lighted half of the moon cannot be seen from Earth.

Hands-On Activity Observe

1. Go outside with an adult on a night when the new moon is in the sky. Locate the moon.

   Where in the sky is the new moon seen?

   ____

2. One week later, go outside with an adult at the same time as you did the week before. Now where is the moon? How much of the moon is lit?

   ____

   ____

3. In another week, go outside with an adult at the same time as you did before. Now where is the moon? How much of the moon is lit?

   ____

4. Draw the changes you saw.

✓Concept Check

1. When you **Sequence** things, you put them in order. See if you can follow the sequence of moon phases. Do the phases of the moon always appear in the same order? Why, or why not?

____

____

2. What happens after the moon rises in the east?

____

____

3. Underline the sentence that tells the ways the moon moves.

4. What causes the moon to appear to move across the sky?

____

The Moon and Earth

Earth has only one moon. The moon is moving, just as Earth is. It rotates on its axis and it orbits Earth. The moon takes about one month to orbit Earth. The moon seems to move across the sky, just as the stars and the sun do. This happens because of Earth's rotation.

The Moon Seems to Change Shape

The moon does not make light. The light you see when you look at the moon is from the sun. Light from the sun reflects off the moon.

Light from the sun is always hitting one half of the moon. The moon looks like a circle when you can see all of the lit half. You can't always see the side that is lit. The moon does not look like a circle when you can't see all of the lit side.

The moon's shape seems to change because the moon orbits Earth. Every night the moon is at a different place in its path around Earth. The moon looks different at each place in the path.

✓Concept Check

1. What happens when light from the sun hits the moon? Complete this graphic organizer to show the sequence of events.

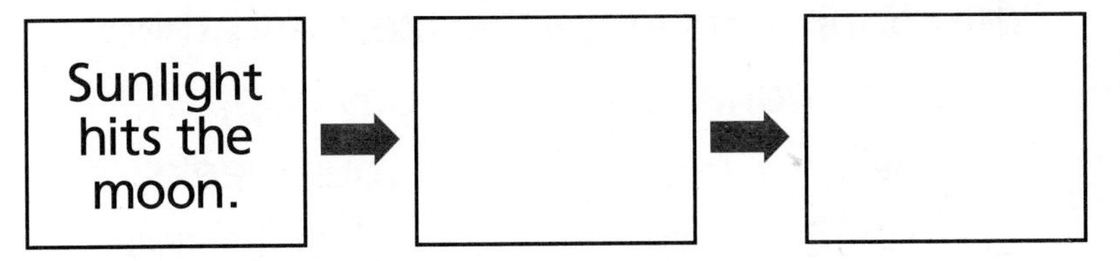

2. The same side of the moon always faces Earth. When does the moon look like a circle?

____

____

3. What happens when the moon moves to a different place in its orbit?

____

____

✓Concept Check

1. When you **Sequence** things, you put them in order. See if you can follow the sequence of moon phases. The lit area of the moon gets larger (waxes) and then smaller (wanes). The changes happen in a cycle that repeats every four weeks. Complete this graphic organizer to show one full cycle of the moon's phases.

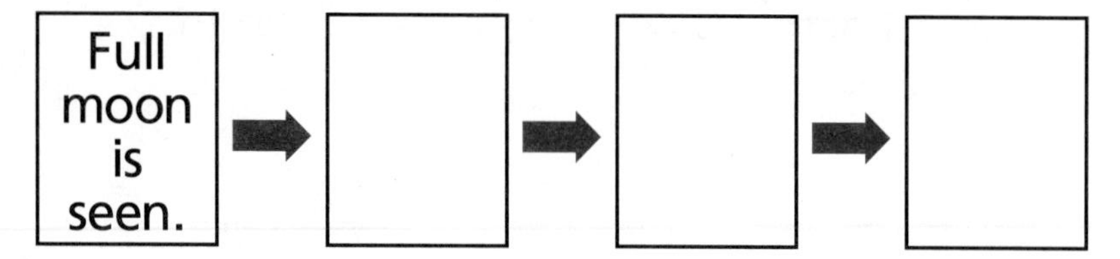

2. What happens after a full moon?

____

____

3. What happens after a new moon?

____

____

Phases of the Moon

The different shapes the moon seems to have are called **moon phases**. Actually, the moon does not change shape. The moon is moving around Earth. From Earth, you see the shape of the moon that is lit by the sun. This shape changes as the moon orbits Earth.

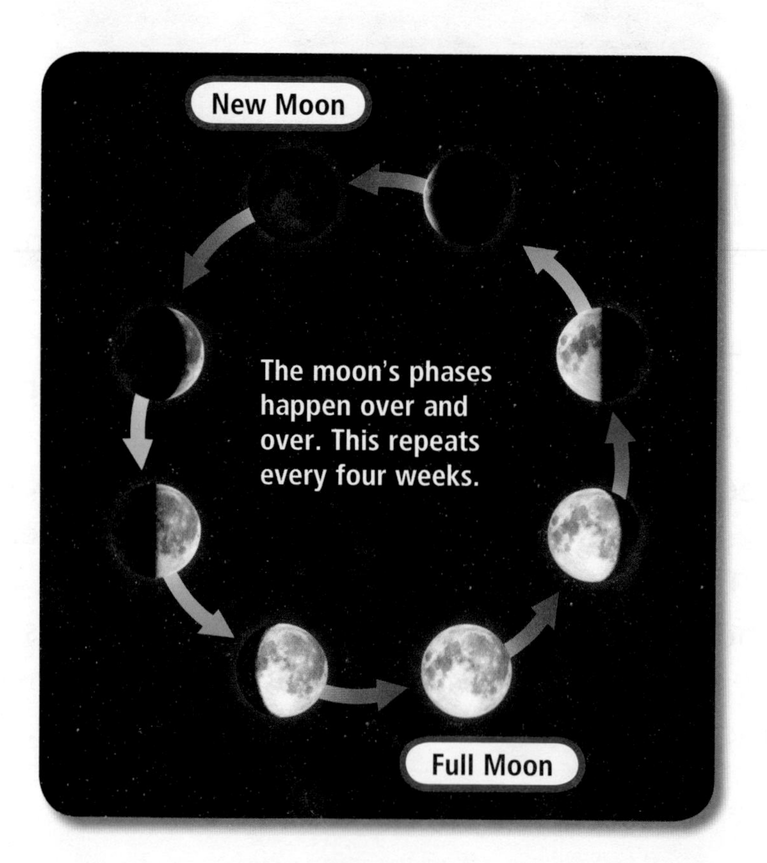

When the moon looks like a full circle, it is called a **full moon**. Then the part that you can see starts *waning*. This means that the moon seems to be getting smaller. In time, you cannot see the moon at all. The lit half faces away from Earth. This phase is called the **new moon**. Next, the moon starts *waxing*. The lit part seems to get larger until it is a full circle. The moon does the same thing again and again.

The calendar shows the different phases of the moon.

Lesson Review

Complete these Sequence sentences.

1. Light from the ____________ reflects off the moon.

2. The ____________ ____________ changes as the moon orbits Earth.

Fill in the missing term.

3. A ____________ ____________ looks like a complete circle.

4. You can't always see the dark ____________ of the moon.

Circle the letter in front of the best choice.

1. Which tells how a planet moves?

   A It moves around other planets.

   B It moves in a straight line.

   C It circles the sun.

   D It circles a moon.

2. Which planet is closest to the sun?

   A Earth

   B Jupiter

   C Mars

   D Mercury

3. How are the inner planets different from the outer planets?

   A They orbit the sun.

   B They are closer to the sun.

   C They are colder.

   D They are made of gases.

4. Which of these causes day and night on Earth?

   A Earth's orbit around the sun

   B Earth's tilt on its axis

   C Earth's spin on its axis

   D Earth's distance from the sun

5. Which word has the same meaning as rotation?

   A spinning

   B orbiting

   C tilting

   D moving

6. Which of these stands for a pattern of stars that always looks the same?

   A sky

   B constellation

   C solar system

   D outer planets

7. Which of these would you need a telescope to see?

   A the moon

   B the sun

   C Neptune

   D the Big Dipper

8. How do objects appear to move across the sky?

   A from east to west

   B from west to east

   C from north to south

   D from south to north

9. How much of the moon's surface is lit by the sun?

   A all

   B half

   C none

   D one-quarter

10. How much of the moon's surface seems to be lit during a new moon?

    A all

    B half

    C none

    D one-quarter

11. You look up at the sky and see a full moon. Which word tells how the moon will be changing for the next several days?

    A rising

    B setting

    C waxing

    D waning

12. List the two ways Earth moves.

    a. ____________________

    b. ____________________

13. Look back at the question you wrote on page 124. Do you have an answer for your question? Tell what you learned that helps you understand the patterns followed by Earth and the sun.

    ____

    ____

    ____